WORDS APPEARING IN *ITALICS* ARE IN THE GLOSSARY

FORWARD.

In a world brimming with uncertainty about our planet's future, it can be daunting to know where to begin tackling the immense challenge of climate change. The noise created is deafening, and it often seems that real solutions are buried under layers of recycled rhetoric and misinformation. That's where this book and its message stand apart.

With a focus on sequestration science, this book reveals the breakthrough discoveries that can transform our approach to combating global warming. It's not just about identifying the problems we face, and all know; it's about offering an actionable solution grounded in undeniable evidence. This book doesn't just inform; it empowers.

Throughout these pages, complex scientific concepts are made accessible through engaging stories and clear examples. This isn't a book for the elite few; it's for everyone who cares about our planet and wants to make the difference. The urgency of our situation is made clear, but so is the hope that only just now is within our grasp.

We now stand at a crossroads. The choice we make today will define the legacy we leave for future generations. Will we continue down a path of complacency, or will we rise to the challenge, armed with this new knowledge and a renewed sense of purpose? This book makes a compelling case for the latter.

Dive in, absorb the logic and wisdom it offers, and join the movement to resurrect the Titans of our forests. It's our last, best hope to beat global warming and secure a livable future for all.

Welcome to the revolution in climate science.
Sincerely, The staff at **MatureTrees.org**

COMPLETE MITIGATION SCIENCE CMS

THE BEGINNING AND THE END OF GLOBAL WARMING ARE ONE IN THE SAME. HOW COULD THEY NOT BE?

Let's dive into the heart of the matter. Excessive atmospheric CO_2 is the unequivocal cause of global warming. However, a typical marketable tree, by age 30, removes 163 lbs. (74 kg) of CO_2 from our atmosphere in one year. By age 70, that tree removes a staggering 1,100 lbs. (498 kg) of CO_2 in one year. There are over three trillion trees on Earth. Unfortunately, 85-95% are under forty-five and stuck in harvest rotations. This means our forests are trapped in immaturity and living within what CMS defined as a *carbon hump*. They are failing to regulate Earth's atmosphere as they were intended. We won't let them.

Now imagine if just 15% of those trees were allowed to mature. The CO_2 sequestration they achieved would erase global warming, surpassing all current CO_2 emissions. This is the power of sequestration science; it is clear, compelling, and currently ignored by those fixated on emission reduction and industrial profit schemes.

The era of ineffective solutions is over. Sequestration science's time has begun. Complete Mitigation Science (CMS) reveals the true climate-changing conditions driving global warming. Sequestration is our most potent tool, way more powerful than emissions reductions alone. CMS studies provide the only known method to completely mitigate and reverse climate change. This requires addressing CO_2 levels in the atmosphere and altering 8,000 years of forestry geo-engineering. That will take time to accomplish, but not as much as you'd think.

Look, global warming is our gravest issue. It's a life-or-death situation, happening faster and it will take longer to cure than previously predicted. Sequestration's ability to measure results highlights that urgency.

All because of this. For millions of years, Earth's CO_2 sequestration cycle worked flawlessly. Today, it's drastically weakened by human geo-engineering. Forest sequestration is now millions of times less effective and still declining.

This sequestration decline leads us towards an extinction-level event, already in motion. Without action, we'll reach a point of no return within 25-35 years. The clock is ticking. This book broadens climate change conversations with groundbreaking knowledge, despite resistance. Information control and ownership obstruct the spread of sequestration's truth. But global warming can no longer hide from this truth. Neither can the fat-cat industrialist. Not anymore.

Recent findings prove comprehensive climate mitigation is now possible, just as the threat of extinction looms closer. We cannot live with global warming's current and future impacts. Sequestration's ongoing decline signals an endless rise in temperatures, which emissions reductions can't stop.

Sequestration science provides the tools we need. Will we use them? This book aims to raise awareness and fund future sequestration efforts. Ignoring these facts guarantees doom and are worthy of your attention.

I now invite you to dive into this book's glossary for fresh, expanded definitions that illuminate the crucial role of sequestration. And thank you for investing in this essential resource. It's vital for raising awareness and propelling future action.

RESURRECT TITANS? SOUNDS DANGEROUS

Let's journey into a mythical tale as old as time. Imagine, if you will, the birth of humanity from the ashes and mist, forged when Zeus cast his mighty thunderbolt, obliterating some of the Titans. And yet he allowed others to live. The Titans, beings who embodied celestial realms like oceans, rivers, planets, and Earth. They were the majestic and cruel predecessors of the Olympian gods. Zeus's cataclysmic war for supremacy not only birthed the Olympians but also sparked the dawn of humanity. At least in poetic myth.

This myth lingers with me, its resonance deepened by the grandeur of the Titans. These ancient entities provided the canvas for all existence. Some were spared to roam the cosmos, while others were imprisoned or obliterated. To me, these fallen Titans symbolize the mysteries that contemporary science so desperately seeks to unravel. Picture Titans named Mathematics, Gravity, Atom, Electron, Quark, and Dark Matter. Perhaps one such Titan was Atmosphere, the keeper of the balance between sequestration and emissions. Ironically, if Atmosphere fell to Zeus's wrath, humanity continues that destruction, annihilating our atmosphere with CO_2 and CH_4.

Yet, in a twist of fate, humanity seized the reins from Zeus with a Titan named Science. Today, our scientific endeavors strive to resurrect the other lost Titans. This is no easy feat; each resurrection brings its own set of advancements and dilemmas. We humans often shy away from the challenges that accompany our leaps in knowledge. Resurrecting this books Titans is about confronting all climate obstacles head-on.

I also invoked the Titans figuratively to emphasize that humanity must embrace new knowledge to advance, while also correcting past misconceptions. Misconceptions like the misguided geo-engineering of our global forests and the idolization of emissions science. These errors, rooted in unverified science, have persisted, and evolved unchecked.

Now, we stand at a pivotal moment. We have the opportunity to craft forestry management plans grounded in Complete Mitigation Science (CMS). These plans can reverse global warming and resurrect billions of immature Titans, restoring their might from the ravages of Zeus and human folly. Understanding CMS, we can declare that resurrecting "Atmosphere" with forestry elixirs is the only cure to restore Earth's sequestration and balance emissions, a harmony that existed long before the advent of mythology or oxygen-breathing life. Thanks to centuries of poor geo-engineering, "Atmosphere" teeters on the brink, as forestry sequestration declines.

I suggest that "Atmosphere" and "Sequestration" are smoldering Titans, desperately needing resurrection by humanity's collective will. This resurrection requires additional dismantling of Zeus's legacy of human domination by improving our stewardship over forestry resources.

To simplify, the immature Titan we must revive today is forest maturity within a portion of Earth's trillions of trees. All because we know the Titan survivors of previous human incompetence and what they do for us. Only now, we know them better than ever.

Today's surviving Titans thrive in a limited number of protected forests, whether hundreds of feet tall or just tens, they can live for thousands of years, growing mightier each year. They absorb CO_2, regulate Earth's atmosphere, cool lands, store carbon, filter our water, produce our oxygen, and truly provide sustenance for all Earth's life. Their growth in rings chronicles their history in the world. All of that only happens if we allow immature trees to become Titans, so they can safeguard our planet against global warming's madness. Add to that, allowing them to grow is no longer an option.

CLIMATE AUTOPSY AND CAUSE OF DEATH

Let's embark on a journey into the heart of Complete Mitigation Science (CMS), a deep dive into sequestration science. To start, we'll stick to the larger facts and zoom out for a cosmic view of Earth; a perspective that stirs the soul and ignites the fight against global warming. A look at something we were given and now stand very close to losing.

From space, Earth is a mesmerizing sight: a vibrant blue marble enveloped in swirling white mists and thicker clouds. This view inspired us to delve deeper into climate science, seeking ways to fight smarter in the losing war against global warming. It offers hope for changing the world, this time with wisdom and for the better. With sequestration, we can achieve a permanence that outlasts our past filled with harmful ideas. We are now capable of protecting the very fabric of our lives, Earth.

Continue to imagine Earth from the vastness of space. The planet's surface is a patchwork of blue oceans, green lands, and brown deserts. These deserts, with their lifeless tan expanses, mark Earth's climate wounds. The scars of drought and growing desertification. Ironically, from this distance view, our bustling cities appear as black voids; deceptive chasms, appearing lifeless yet teeming with human activity. The blacken voids and scars of tan remind us of our responsibility to use Earth's resources wisely, to preserve our fragile habitat, our only home and also to advance.

As night falls, those black voids transform into dazzling constellations of amber and yellow lights, a testament to human brilliance. It's a spectacle, like a grand Disneyland fireworks show, highlighting the contrast between our daytime blacken void with the display of nighttime vibrancy. This striking contrast highlights humanity; our ingenuity and also our advancing impact on Earth. And yet, that celestial view also serves as a stark reminder of our tiny place in the vast, cold cosmos. It emphasizes our true dependence on Earth, our only home, a fragile oasis in the dark and frozen universe. A reminder of our survival that hinges on our ability to care for this planet. Earth will always endure what may come; it is our future that's at stake. We must become wise enough to honor and protect our only home. Yet we only just now hear Earth's warnings. But Earth's message is clear: act now or face dire consequences. CMS hears that callous cry and now, you will.

Through the lens of CMS, we've gained the knowledge and tools to make a profound and lasting positive impact. Sequestration science offers a beacon of hope, guiding us towards a more sustainable future. By embracing sequestration science, we can restore the balance of sequestration and emissions and heal Earth's wounds. Wounds we've unintentionally inflicted but are entirely accountable for.

Earth's survival is guaranteed, but ours is not. To ensure our future, we must embrace our role provided by Earth. We must be better stewards of Earth, our only home. Make no mistakes about it, fixing Earth is now our one and only job. The time for action is now because there is no saying sorry to Earth tomorrow and all will be forgiven. There is no tomorrow, not if we don't change our sequestration ways. So, I've written this book to explain our ways. A book that will take you on a journey through sequestrations groundbreaking discoveries, showing how we can turn the tide and secure a livable future for generations to come. All because…

WE MISSED SOMETHING ALONG THE WAY. SOMETHING IMPORTANT. SOMETHING THAT SEQUESTRATION'S ONGOING DEATH IS REVEALING.

A few hundred years ago, Earth's forests were vast and thriving, absorbing far more CO_2 than they do today. These ancient forests were incredibly efficient at sequestering CO_2, a process essential for maintaining the planet's climate balance. The trees, some of which were centuries old, played a crucial role in the natural carbon cycle. Also known as the plant growth cycle and as the fast-cycle CO_2 or carbon sink cycle.

Back then, these forests could absorb more CO_2 than the Earth produced naturally, or humankind could ever release as emissions. This immense capacity for carbon sequestration meant that global warming was literally impossible. The forests acted as giant sponges, soaking up CO_2 and keeping the atmosphere in check.

However, around 250 years ago, things began to change. Human forest activities began todays pinnacle as they started to alter these forests significantly. The Industrial Revolution marked the beginning of large-scale deforestation and land-use changes globally. Forests were cleared for agriculture, timber, and urban development. This transformation also reduced the remaining forests' ability to absorb CO_2, with *demand driven forestry* and *constrained deforestation*. Which lead to an increase in atmospheric CO_2 ppm (parts per million) levels.

CMS has been pursuing sequestration-based science to understand this correlation. The research has shown a direct link between human forestry practices and the rise in atmospheric CO_2 levels. This correlation revealed global warming was a well-hidden environmental impact. It was found within the study's well defined forestry activities and models.

The reality we must embrace is that forestry photosynthesis historically could absorb more than all combined emission sources could ever emit, up until about 250-300 years ago. This means that before significant human intervention, the Earth's forests were capable of maintaining a stable climate by absorbing CO_2 before it became an excessive level within Earth's atmosphere.

Today, the situation is different. The excessive levels of CO_2 in the atmosphere are estimated at 1,494 gigatonnes as of July 2024. That level is the major driver of global warming's effects. This level of CO_2 is increasing because of what Earth's current, diminished forests can't absorb, but had in the past, and could do again after mitigation.

The importance of mature forests in regulating CO_2 levels cannot be overstated. Mature trees, which have been growing for decades or even centuries, are incredibly efficient at sequestering carbon. Expanding on the earlier example, a tree that is 30 years old can absorb up to 163 pounds (74 kg) of CO_2 annually. By the time it reaches 70 years old, it can absorb around 1,100 pounds (498 kg) of CO_2 each year. This capacity continues to increase as the tree ages.

Trees increase their CO_2 absorption capacity every year they are allowed to mature. Typically, this increase ranges from 3-15% annually. Some tree species can live for hundreds or even thousands of years, continually enhancing their ability to sequester more and more carbon throughout their lifetime.

The degradation of tees and forests has had a profound impact on the Earth's climate. Without the vast, mature forests of the past, CO_2 levels have risen unchecked, leading to global warming. This warming effect is not just a result of emissions but also a consequence of the loss and ongoing impediment of these critical carbon sinks.

To address this issue, the protection of existing forests is essential. Protecting existing forests from further degradation and promoting their maturity is important for them to maintain their role in regulating the Earth's climate. This book explains how.

In conclusion, the story of Earth's forests is a reminder of the delicate balance that once existed and the impact of human activities on this balance. By understanding the importance of mature forests and taking action to restore and protect them, we can mitigate global warming and ensure a stable climate for future generations. Doing so avoids Earth's alien autopsy, like the ones our own planetary science perform on the dead planets in our solar system.

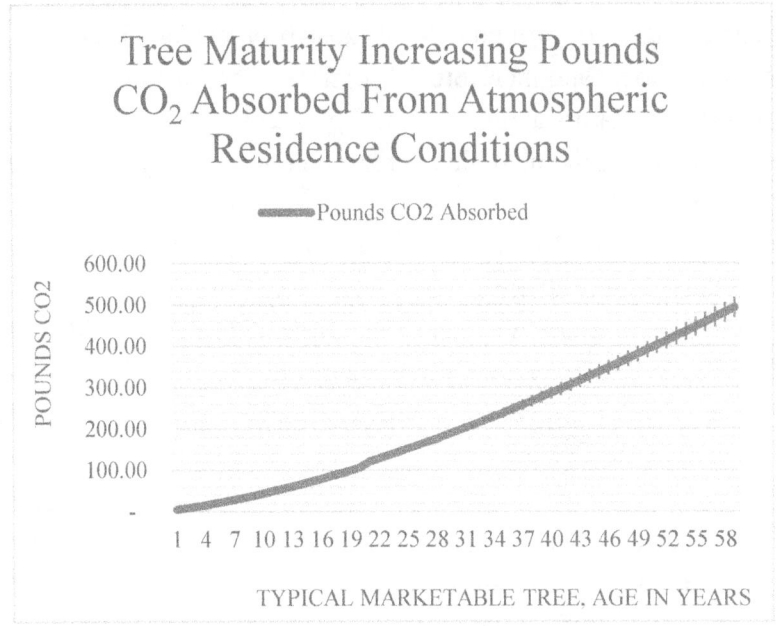

CMS REVERSES GLOBAL WARMING WITH MATURE TREES – THE TITANS OF THE FOREST. SEQUESTRATION SCIENCE EXPLAINED HOW.

Figure A, graphing maturities significant impact on forestry CO_2 sequestration. Tree growth's effect on atmospheric CO_2 sequestration is proven for 59 years of maturity. The tree used in this demonstration can live over a thousand years. [i]

Again, and to expand, according to the United Nations and other sources, it's possible that **three trillion trees exist here on Earth**. More trees exist than stars. How about that? We certainly have plenty of trees to work with.

Unfortunately, and as I've mentioned, most of those trees, or at least somewhere between 85-95% of them, are "marketable" and therefore subject to harvesting. The inhabitants of Earth need the stuff we make from trees and that requires all forest lands to basically support that economy. So, only 5-15% of forest land globally (the UN says 15%) is protected from harvesting while around 80% of those protected areas are found suspect and therefore susceptible to harvest given the mood of dictators and the possibility of corruption involved in not protecting them. Which is reason enough to say 5-15% instead of 15%.

For example, Russia is the single largest tree producer in the world with over a trillion trees. Then again, we should all understand current evidence. Dictators and Oligarchy did not protect forests, they annihilated them for cash. Within those types of countries, you'll be fortunate to find 30-year-old marketable trees.

Globally, all the forest land available has been harvested and now seriously lacks maturity. Thats the point I need people to understand. Forest maturity is spent from the entire globe's forests. Very little remains.

Mature trees (Titans) are indeed a global rarity. Furthermore, most of those forests have been harvested many times and suffer *tree and land degradation*. Good luck finding a tree anywhere on "unprotected" land that has matured past forty-five. Or one anywhere that achieves its genic abilities due to *land degradation*'s lack of nutrients. Most global trees are under 45 years of age and the average age and size is declining fast due to *demand driven forestry* and *tree and land degradation*. Both are *constrained deforestation*. To make that fact worse, most of the Earth's Forest lands are devoid of trees due to harvesting or filled with too many immature trees waiting for lightning or a careless matchstick to ignite them.

Globally, clearcut lands mostly exist as non-replanted lands stripped of trees for decades, or they are in "natural regeneration" with only a few reseeded sprouts or saplings covering small sections. Natural regeneration really means not replanted, which isn't necessarily a bad thing to anything but maturity. Which makes it bad for <u>everything</u>. But it is worse than just maturity.

I do hate making anything worse, but I must. The United Nations acknowledges that up to 30-40% of Earth's previously forested land has disappeared for different land uses. AKA, land use modification. Most of which disappeared in the last 250-300 years. And all of which fell to mankind's undertakings. The loss of forest due to land use changes combined with *constrained deforestation* are geo-engineering practices responsible for climate changing conditions and cause global warming. The bad part of that is that forests disappearing globally outpace our afforestation practices and does so by extremes. We are losing forests to inefficiency and greed.

For real time proof, it's time to jump into <u>Google maps</u> and use their satellite imagery to take a tour of the entire world from space.

Spending a few minutes or a few hours looking at the world's forests says it all. Notice the lack of trees on harvested grounds in "natural regeneration." How about the sprinkles of immature trees that look like grass, most of which can't even be seen due to Google's low-resolution satellites. You will see that even the large trees are less than 45-60 years old. Just compare them to a US national park (with trees) or non-harvested Amazon Forest region to see the difference. And now I can offer my congratulations, you just made another CMS proof. What you proved with your own view is what this book is all about. Our environmental impact on forestry; our undesired geo-engineering outcome; human efforts that caused and continue to make climate changing conditions. More importantly, how we can now define it.

This book tells of twisted facts that you're not going to like, at all. But I'm going to tell you anyway, whether you, me or anyone likes it or not. I'm told I'm like that. I made a career out of telling people what they don't want to hear. I call that grandpa's curse and explain later. For now, I'll admit that I don't have a monopoly on all the good ideas and don't mind feedback or peer review. Plus, I like sharing sequestration science's news that provides hope for the future. Now, how do I say this next part? Well, this book is intended to provide hope by adding some light to the darkness surrounding global warming. Darkness, I am also forced to share.

THE TIME FOR ARROGANCE HAS ARRIVED.

Let's make one thing clear: reducing human emissions is essential. There are countless reasons to do so. But let's face it, emission reductions alone are not a climate cure. This isn't just my opinion; it's backed by hard science. Sequestration science doesn't just talk; it delivers results we can measure in our atmosphere and in our forests. It provides concrete proof, backed by multiple methods and peer-reviewed studies. Emission reductions? They're based on assumptions and their not measurable without those assumptions.

Reviewers of Complete Mitigation Science (CMS) can easily reproduce or build their own sequestration models. Even with existing studies, many of which predate CMS by decades. Studies, I had nothing to do with. Adding up all the open-source and verified studies also confirm the undeniable truth of CMS. So, this book isn't just nonfiction; it's a highly referenceable guide to our climate reality, Titans metaphor aside.

Let's confront the tougher facts. Emission reduction science falls short. It cannot offer the concrete, climate-impacting proofs that sequestration science can. It can't validate its predictions or show atmospheric results implied. Over the past 30-50 years, emission reductions haven't significantly slowed or decreased atmospheric CO_2 levels. It's time to question their effectiveness. Meanwhile, sequestration science directly addresses both natural and human emissions while it impacts CO_2 excesses in our current atmosphere.

Here's why I can say all that with confidence: each year, 300-750 gigatonnes of CO_2 are emitted naturally. Emission reduction science has no control over these emissions, but sequestration science can manage both natural and human emissions. Moreover, there's 1,494 gigatonnes of excess CO_2 in Earth's atmosphere that emission reductions can't touch—but sequestration can.

People often argue that increased atmospheric CO_2 levels are due to rising human emissions. Post-CMS, this argument simply echoes very outdated findings. We know now human emissions are only 8-10% of natural emissions levels. No doubt human emissions are too high and thanks to the CMS carbon hump higher than previously believed; however, they've proven to be unavoidable. Humans are *emissions dependent* for survival. Natural emissions, which have been around before humans, are entirely unavoidable.

In the ongoing struggle of human self-domestication-our quest to master our environment-CMS proved that the rise in atmospheric CO_2 parts per million (ppm) is due to the decreased sequestration ability of our forests. This decline in sequestration is linked to the lack of mature forests necessary for regulating CO_2 levels.

The real cause of climate change is an environmental impact, not emission levels. It wasn't the industrial revolution, increased emissions, or fossil fuel use that initiated climate change. Oh, I know, that is really hard to believe but now proven true. A breakthrough even. All those factors are merely variables within the much larger issue of global warming.

To sum it up: the increase in atmospheric CO_2 is not just about emissions; it's about Earth's dwindling sequestration capacity. Sequestration science addresses this head-on, providing a path to genuinely combat global warming without making assumptions. It's time to embrace this truth and act accordingly. It's time to Resurrect Titans. To do that. There is a process that requires more than just time.

NOW AND THEN IS NO DIFFERENT TODAY.

I've often wondered if societies from 5,000 or even 10,000 years ago were fundamentally different from ours today. Imagine taking a society from 5,000 years ago and placing them in today's world; how would they fare? Fundamentally, those ancient individuals are no different from us. The only true difference is the tools we use and how we govern our behavior with some of those tools. Teach them our modern tools and expectations, and they would become laborers, airline pilots, professors, doctors, and politicians, just like us. Human brains haven't evolved significantly in the last 5,000 years. Whether viewed scientifically or biblically, we are the same people. However, 5,000 years of self-domestication has advanced our science and technology, allowing us to create better tools. Tools are the key component in our self-domestication.

Tools represent mankind's progress in knowledge, not changes in how we think or instinctively react. Governments, education, and technology are proof of our self-domestication. Eons of hindsight and accumulated advancements have made our modern tools and problems possible, and we will continue refining our tools faster than our brains can adjust.

The Dunning-Kruger effect explains our limited understanding and overestimation of our abilities. Are we really that dumb? Not exactly. But CMS has revealed that we knew less about climate change than our self-proclaimed expertise allows. Emission reduction experts deserve credit for their efforts, but CMS's sequestration science is redefining the field. CMS forces us to confront our misunderstandings and move forward with a much better tool, sequestration science.

Historically, every significant push in knowledge faced persecution before its achievements were recognized. Imagine the first cave dweller lighting a fire, only to have it extinguished out of fear. CMS is the first to ignite sequestration's fire, expecting opposition but welcoming peers. Let's hope the opposition fades before it extinguishes this crucial spark.

Good science should be our self-domestication foundation, with facts as our guiding truths. But that is not what happens. Communicating good science is difficult at best. To make this good science more accessible, I've written this "shiny" book that has been called arrogant. CMS isn't about arrogance; it's about documented science that warns of the consequences if we don't act quickly with sequestration tools. Timeliness is crucial and a motivation to communicate it as best I can. Time is also related as some climate change events are set in stone, like atmospheric CO_2 levels from 1972, while others, like volcanic eruptions, could unfold at any time.

Mother Nature's message is clear: act now or face dire consequences. 5,000 years ago, people would evaluate CMS just as we do today. Societal acceptance of new knowledge remains a challenge, regardless of its proven benefits. Decisions not based on sound science are prone to failure. Scientific advancements offer clarity, unlike the instinctive and emotional judgments we often replace it with.

CMS is a modernized understanding of old principles. It integrates fundamentals in ways the highly specialized world has overlooked. It's interdisciplinary, a Jack of all trades, and challenging to relate to the uninitiated.

To be sure, we all recognize the ongoing climate war we're losing. CMS is the first to explain why we're losing. And we can see the impending disaster if we don't act. Together, we can shift from assumed knowledge to certainty. Today, the biggest battle in the climate war is the information battle, where sequestration science struggles to gain traction. Listening and acting on CMS is the only way to secure a climate victory for humanity. There is no other option.

EARTH'S TO-DO LIST:

[1] ~~Reducing Emissions Will Cure Climate Change! Nope, but good for many other reasons~~
[2] ~~Solar Panels Will Cure Climate Change! Nope, stepping in the right direction~~
[3] ~~Battery-Powered Cars Will Cure Climate Change! Nope, they add to it~~
[4] ~~Wind Power Will Cure Climate Change! Nope~~
[5] ~~Nuclear Power, That'll Fix It! Helps, probably worth the risk, doesn't cure climate though.~~
[6] ~~Tiny Houses! We'll All Live Like Mice! Helps, but doesn't cure climate~~
[7] ~~Reduce, Reuse, and Recycle Will Cure Climate Change! Significant effort! Keep it up and no.~~
[8] ~~Planting Trees Will Cure Climate Change! KIND OF SORT OF!~~
[9] Cap and Trade Will Cure Climate Change! It helps now and can help more!
[10] ~~Technology! We'll Invent Something to Cure It! Not Likely~~
[11] ~~Humans Can All Go Back to Being Hunter-Gatherers! And Billions Starve to Death~~
[12] ~~Let's Become Vegans, That'll Cure Climate Change! It won't help, it makes it worse~~
[13] ~~Off-Grid Living Will Cure Climate Change! Helps, but doesn't stop climate from changing~~
[14] ~~Our Government Will Cure Climate Change! We The People Can Make That Happen~~
[15] ~~Reduce the global population! Nope, that isn't required, not yet~~
[16] ~~Climate Change Isn't Real! YES Virginia, there really are climate-changing conditions that create global warming!~~
[17] Complete Mitigation Science? Finally, an actual cure for climate change! You can see, touch, and smell it for yourself!

WE'VE BEEN THERE AND DONE THAT, AND NOTHING WORKED. BUT DID WE TRY EVERYTHING AVAILABLE. NO, WE MISSED SOMETHING, SOMETHING IMPORTANT. WE DIDN'T USE SEQUESTRATION SCIENCE BECAUSE WE DIDN'T UNDERSTAND HOW IT COULD WORK. BUT NOW WE DO!

Regardless of CO_2 emission's sequestration science has the cure for all of Earth's CO_2 excesses. Those same CO_2 excesses contributing to global warming and our shared fate. We are *emissions dependent*. Except now global warming is no longer existential because CMS studies proved global warming causes and its ongoing resiliency that plagues Earth.

CMS's method offers the only way tested and proven to actually cure global warming permanently. No kidding, and no ifs, ands, or buts to it. A known cure. It works because it had for eons, but we broke it, and I mean humans demolished it. Anyway, CMS actually addresses all documented sources of CO_2 emissions by fixing what used to be very abundant, is a naturally occurring fixture, and undoubtedly the most inexpensive and beneficial way possible, Titans.

Prior to CMS, I was literally just like the lined out but very admirable and useful ideas on that "To Do" list. I was, like many of you, completely dedicated to all those listed concepts' promises. Unfortunately for my climate snob ego, CMS pointed out they do little to nothing in actually curing climate change. Crueler yet, some of them add to our climate crisis. And those particular but well-proven CMS facts can be extremely hard to swallow.

Many of those lined out efforts improve our environment, restore our biomes, and some of them improve our overall well-being. They can also provide abilities for better living while sharing with others. So, they can address other problems just not climate change directly. Indirectly, some do; but CMS still has to line them out of the way because they can't and don't cure climate change at all, not even when combined.

Please keep up the pursuit of our progression in bettering how we live, even if CMS lined it out. Just remember, they won't stop, slow down, or cure global warming. Without CMS involved on their sequestration side, they can't. But nobody understood that when they were developed, nor when their developers told everyone to treat emissions reductions as the only way. The people who developed them took the first steps, sequestration science has ran past them. We've advanced.

FORESHADOWING

A **skip** forward to "To Start, Try Explaining CMS To Anyone" begins the more detailed and proof sections.

For now, I will continue foreshadowing the importance of CMS without diving into much detail. By highlighting the significance of CMS's global forestry sequestration model and its methods to repair what took humans eons to damage, I aim to offer a concise yet comprehensive summary to accelerate the dissemination of knowledge. Here's the essence:

CMS is built on the understanding that mature forests play a critical role in regulating our climate through sequestration. Over centuries, human activities have severely compromised this natural system. CMS provides a scientifically-backed method to reverse-engineer our climate's changing conditions by restoring the maturity and health of our forests. This approach not only targets the root cause of global warming but also offers a sustainable, economically viable solution that integrates seamlessly with our future forestry use. By leveraging the power of sequestration, CMS aims to create a more stable and livable environment for future generations.

To encapsulate the main features of CMS succinctly:

[1] **Global Forestry Sequestration Model**: Emphasizes the restoration of mature forests to regulate atmospheric CO_2.s
[2] **Reversal of Climate Change**: Uses scientific methods to reverse-engineer climate changing conditions.
[3] **Sustainable and Economic**: Integrates with current technological advancements without sacrificing economics.
[4] **Long-term Solution**: Targets the root cause of global warming for a sustainable future.

THIS OVERVIEW SETS THE STAGE FOR A DEEPER EXPLORATION OF CMS, PROVIDING A FOUNDATION TO BUILD UPON AS WE DELVE INTO THE SPECIFICS.

HARVESTING TREES THE RIGHT WAY.

Yes, CMS advocates harvesting trees because economics dictates that is required. Just as much as increasing *sequestration value* is. CMS only promotes harvesting under natural attrition's strict criteria. *Natural attrition harvesting* is nothing like the way forests are currently harvested or managed. Natural attrition methods only select dead and dying trees. Natural attrition's criteria mean's maturity first, which provides gains in sequestration, and biomass efficacy. The process allows more biomass to grow and does so without clearcutting or sacrificing maturity.

Maturing forests, beyond todays harvest intervals, grow incredible amounts of dying biomass while natural attrition criteria manage them. Which is the opposite of what happens today. CMS is also focused on using natural attrition to produce economic benefits. It needs to be done or sequestration improvements with maturity could die on the vine. But the real reason is *natural attrition harvesting* creates far more sequestration and biomass efficiency than today's methods do. It also significantly decreases immature forest *carbon humps*. Unfortunately, it also requires patience, initially.

HINDSIGHT CAN BE LEARNING, THE BEST LEARNING.

The human mindset, specifically changing it from bad to good, requires foresight and hindsight. Using hindsight to get better facts through earlier studies works well. The trial-and-error method. So, CMS researched emission reduction failures that eventually pointed out the way to end climate change, with sequestration. Hindsight delivered a message that said we are missing something in climate change's understanding. Hindsight might be the best human tool we have for science. It sure helped isolate facts for CMS to point out. These hindsight comments are meant to impart learning from our mistakes as a good thing.

Learning in hindsight offered CMS ways to make big steps forward in climate mitigation, climate predictions, atmospheric science, and forestry. It's not $E=MC^2$ it's sequestration greater than or equal to emissions and something we can all unfortunately see and feel the results of. And I'm not talking about seeing and feeling climate change results, I'm talking about seeing, touching, and smelling climate changing conditions because CMS can actually point to what causes it all. Climate changing conditions had hidden in hindsight, until CMS put the spotlight on them so they could admit their atrocities.

CMS defines the knowledge of climate change's origin and persistence. That came unexpectedly. Without hindsight, it is impossible to prove. Climate change's origin and persistence wholly surprised me and left me speechless. That's because it is CO_2 emissions related, as current perception believes, but not anywhere close to the emissions efforts we all subscribed our souls towards. And bet our lives on.

This book is my best effort to spread CMS's improved climate change knowledge everywhere. I feel that I must, because this book relates factually to cause, effect, impact, and ultimately defines the only cure for climate change possible.

My intent in authoring this book is to provide knowledge, warning, and hope. And it must do hope; because hope is all we have left to work with. I'll explain that later in detail. For now, CMS changed my understanding of climate change from it's no big deal to ah-crap we really are doomed and then into or are we. Oh, and a second ah-crap when I learned it's coming faster than previously predicted or currently believed. So, we now know if we ignore sequestration science and don't apply hope, we aren't going to be talking about a two-degree United Nations predicted and accepted temperature increase. Try more like 7-10 degrees in the same predicted timeline. Yeah, it really is that ugly and even the IPCC (Intergovernmental Panel on Climate Change) expert predictions are found too cold as their predictions come into fruition! So, they have been and will be wrong again. The IPCC is still without sequestration computation which explains the too cold predictions. Well, if I didn't make my point let me be forward. Hindsight and CMS can explain why they've been wrong and can help make existing prediction methods more accurate.

Here's the thing. Emission based climate science everywhere is fundamentally incomplete and not reproducible without CMS's sequestration-based computation helping emission-based science to form a completed forecast. Sequestration's analysis of current emission reduction efforts provides a better foresight. All because emissions reductions are devoid of accounting for their hindsight. Which leads us to you and yours.

You And Yours.

Sequestration is something we all need to empirically understand in order to end the single greatest threat humans have ever faced. The reason sequestration understanding requires each of us is because climate changing conditions required millennia's to form and not the decades or centuries emission's science would have you believe. It took human hands over 8,000 years to geoengineer global warming into today's single largest problem. Literally, tens of thousands of generations before us created our climate problems today with their lack of knowledge influencing their forestry decisions with coins and not sequestrations foresight. That influence born a negative and still growing environmental impact. Hindsight again helps us to see that our predecessors and us didn't understand what was happening to Earth's atmosphere with greed's globalized forestry practices; but, thanks to CMS we do now. Only now do we understand that generations of our predecessors broke sequestration, so now it's going to take all of us living decades to fix it; and our descendants to maintain it. In addition, it now seems our lack of dedication to forestry health is no longer the option we believed it was.

A CLIMATE OGRE? YES, BECAUSE THE LORAX IS A WIMP

The Lorax's warnings fell on deaf ears, and only after the trees were gone did people listen. Regret after the fact is how criminals justify their actions. But being sorry doesn't fix anything. Embracing the Lorax's wisdom when it's timely is a human trait we all possess, yet not everyone utilizes.

Real life isn't a Dr. Seuss fairytale, and the Lorax wasn't persistent enough. He missed a crucial point: no tree maturity means no humans. It's that simple. Today, we face countless problems. *Demand-driven forestry* and *land degradation* are key contributors, causing catastrophic forest losses. And totally ignored by climate scientists because they are sequestration related, not emissions reductions.

Climate change discussions are hijacked by profit motives, political agendas, and a false sense of security in facing predicted repercussions. Many have given up the climate fight, and who can blame them? We've been misled and deceived so often that it's no wonder we're losing the war against global warming.

At the time of writing and in the public, any climate change cure seems impossible, given past emission sacrifices and expenses. But there's hope! Sequestration science eradicates this hopelessness. With sequestration's advancements, we can now measure, forecast, and fix the climate crisis. No more emissions-based guesswork. We've learned from our mistakes, making us wise like the Lorax, but stronger-like an Ogre.

Fairness dictates acknowledging that those making climate decisions are doing their best with what they know. They currently lack sequestration knowledge. Hopefully, they heed the Lorax's warnings before the climate Ogre gets involved.

Here's the crux: we can end climate change, but it might require drastic measures; rallying the village with pitchforks and torches. The world's leading authorities on climate change and the wood industry don't yet understand sequestration math or tree maturity. Their current mindset isn't born out of malice but misunderstanding. It's unfortunate, but their beliefs may trap them in ineffective actions. However, the time for gentle warnings is over. Now, we need to become the Ogre, wielding a knotted club and shield of truth. Uninformed people must become informed whether they like it or not.

THE HARSH TRUTH ABOUT SEQUESTRATION AND GLOBAL FORESTRY

The CMS study reveals a stark reality: while sequestration is trying to regulate our global climate, it's currently in a dangerous state of decline. This retrograde condition is largely due to human activities, which have created many accumulating declines, all are linked to forestry. What remains of our global sequestration efforts is severely hindered or trapped within the carbon hump, having minimal or no positive atmospheric effect.

However, there is a glimmer of hope. The few remaining "Titans"-old-growth, mature forests, and scattered mature trees are still making a significant atmospheric impact. These Titans, though rare, continue to work hard for us. They are majestic and critical in their role. The rest of the trees, unfortunately, are ineffective in terms of atmospheric impact. They are Titans in need of resurrection.

Historically, mature forests have played a crucial role in regulating Earth's climate for hundreds of millions of years. Yet today, their effectiveness has drastically diminished. Forestry sequestration is in failure mode, with less than 15% of global forests showing any measurable maturity. This lack of mature forests, which previously balanced CO_2 levels, poses a significant threat to our survival.

The majority of global trees are stuck in the *carbon hump*, have been eradicated, or are too immature to make a difference. These immature trees are actually contributing to emissions rather than reducing them. It's a hard pill to swallow, but today's forestry is far from green.

WE ARE SEQUESTRATION STARVING PEOPLE. AND LITERALLY CUT DOWN OUR TREE FOR FIREWOOD TO KEEP IT FROM PRODUCING IT'S FRUIT.

The global sequestration failure is happening because forests are only managed by the economic return found in *demand driven forestry*. That negates the second renewable CMS found within trees, their *binary restricted resource*. That forgotten renewable is sequestration as *sequestration value*. The trees ability to remove CO_2 from the atmosphere. Forest management schemes broke sequestration abilities and limit them by valuing the faster renewing and profit driven immature trees over the better managed and better producing mature trees that renew sequestration and biomass faster. It's a timing thing that humans implemented to cash in right now off immature trees rather than wait for the bigger and better mature trees. Economic impatience is the problem, as I define it.

We broke global forest sequestration while pursuing forestry use. How it continues is a shock to me, although saying that is hindsight to CMS. Yet, destruction created without foresight seems to be engrained into human nature, our first instinct is to gain and then suffer the consequences in lack of foresight.

Discovering one of the impacts of what CMS coined as *demand driven forestry* provides many valid points towards suffering our gains. One such point is made historically, when around *1850ish a Climate Datum* appeared as the atmospheric CO_2 scale tipped away from a symbiotic environment to a CO_2 driven catastrophe in the making. Catastrophes that continue to unfold include decades of drought, off-season record-smashing storms, and the record-breaking heat waves we experience today.

TOO YOUNG TO KNOW

I grew up watching logging and forest management practices. But I had never pictured those same exact activities and the same "bad" forest management schemes happening simultaneously across the entire globe. Literally, there is little to no forest on Earth that has not been logged at least once and others as *convenient forestry* have been logged hundreds of times. Ever since humans learned how to use forest resources we've been logging. Literally, every forest on Earth is in a harvest rotation or considered a strategic reserve.

In fairness, some western forests in the USA were cut for the first time in the 1980's-1990. Only because they could be. It happened when the price of timber increased enough to justify the expensive roads to access the last old-growth resources. Tragically, you can see after they were harvested the atmospheric CO_2 ppm (parts per million) increase. Those last old-growths are found wanted today. But we cut them down for miniscule economic gains for a few already wealthy timber barons who depleted their own timber growing lands. Their substantial political influence gave them access to even the most remote old-growth stands in America. And still does.

For the record, using forestry resources is fine and dandy, but bad forest management is not. Here in America, we have to learn how to say no to one sided decision. The way we manage forests must prioritize the return to a mature forestry norm before any harvest is considered. In other words, mature, living trees must not be harvested. Natural attrition harvesting must take the place of contemporary harvesting. That also improves the amount of biomass that can be harvested in the future. So why don't they do that? Well, they haven't learned the benefits of CMS yet, maybe? I can believe that, somewhat. But the economic impatience prevalent is what it is-greed, plain and simple.

DON'T OVERCOMPLICATE THE PROBLEM

Let's keep it simple: science has overcomplicated the climate issue with emission reductions, and by going along with it, we've created an information mess that CMS needs to fix. For CMS to succeed, it's crucial to understand the differences between emissions and sequestration methods. Investing in sequestration science, the proven cure for global warming, requires a significant shift in global perception, which is easier said than done.

CMS is the out-of-the-box approach, and rogue. Before CMS, no one considered that forestry data could connect atmospheric CO_2 levels to historical forestry declines or recognize forestry's impact on global sequestration and our climate-changing conditions as a cause. These correlations are key to understanding how human-engineered environmental impacts impeded sequestration.

To relate this personally: we were all wrong about climate change's origin and impact, and deep down, we knew it. We avoided admitting our climate losses or discussing looming failure because survival is our goal. We now see that the fight was misled by emissions science, which couldn't definitively explain why atmospheric CO_2 levels kept rising. CMS disarms both skeptics and emission reduction science by providing reproducible results, proving CO_2-driven global warming is a significant issue and that reducing emissions is a variable, not a cure.

Before CMS, all we could prove was that climate change is winning. Now, sequestration science challenges everything we thought we knew. It has made all of us look dumb. So, we're reluctant to change our perceptions; but we also know we must. Proving the correct solution means letting go of longstanding beliefs and past investments. Sequestration's knowledge is undeniable; it's the true prescription for curing global warming. The only one proven to actually work.

It's time to embrace this truth and revive the sequestering Titans who have long supported humanity's science. Recognizing sequestration science's benefits makes a timely weapon against global warming, it is ready to lead the charge.

THAT ENDS MY SHORT VERSION OF CMS.

From here on, I'll be adding more points and details to our conversation. I'll introduce a bit of math, chemistry, and other disciplines to explain and prove CMS concepts. I'm not claiming to be unique in my understanding or expertise. My aim is to write about CMS in a way that everyone can grasp. Your patience is appreciated as I occasionally veer off on related tangents. I might even complain a little about how we got here in the first place.

Keeping this book interesting was a challenge due to the dry nature of CMS's facts. As an engineer and technical writer, making sequestration points poetic was difficult. I'm not much of a cheerleader, but I do have some compelling advantages to share. CMS's facts are compelling, and it's a save-the-world kind of thing, so it should be communicated effectively.

So, I will guide us along this hot, dusty path and steer us clear of the abyss. But you may want to leave the light on tonight. It's going to get a little scary.

TO START: TRY EXPLAINING "CMS" TO ANYONE

First, I think I should start with some principles. They'll shore up the logic and proofs coming.

GLOBAL AVERAGE TEMPERATURE.

A global average temperature fundamental is temperature is found to be inversely square to CO_2 ppm levels and not directly *proportional* to ppm level's increasing or decreasing. As a positive feedback loop heat creates more heat or freezing creates more freezing. Either of which increases or lowers temperature significantly faster than CO_2 ppm levels increasing or decreasing. Therefore, increasing ppm by any amount accelerates temperature increases faster and faster as temperature approaches its apex. The apex here is the highest achievable temperature; given all the variables that could affect it. For example: The planet Venus's temperature apex is an average temperature of 860° F (460°C) which by its own example could also be something like Earth's temperature apex. Let's not find out.

Temperatures ability to accelerate is an indicator that clearly defines global warming's ability to wipe us out very quickly. Additionally, science has recorded quickly accelerating global average temperatures as proof. That is not assumed, it is very real. That acceleration is understood and defined as one of the accelerating components of *accumulating declines*. It is one of the what's in the, "what needs to stop" question to cure global warming.

Okay, I regretfully will use a cliché, we are the happily swimming frog in a pot being heated to a boil. For the record, 212° F, 100°C boils us. Here on Earth and when we are enduring heat waves, we're more than halfway there already.

CARBON, C, AND CARBON DIOXIDE, CO_2, RELATIONS.

We all recognize CO_2 as carbon dioxide, that vile molecule within our emissions and the stuff accumulating in our atmosphere that is without doubt causing global warming. Well, it really isn't vile. CO_2 generates life in more abundance than the O_2, as oxygen, that animals breathe. That is, when human's let it. Not letting it is a cause of global warming.

A carbon atom is well known as having twelve atomic mass units (amu) or some even refer to it as C-12, carbon-twelve, for short. Twelve amu is carbon's mass. So, twelve amu will accurately model the amount of carbon stored within a tree or really anything and everything by mass. However, isolating only carbons mass (just under half a tree's total dry weight) misses the big atmospheric picture for a couple of reasons. First, because of one question. Where did all those carbon atoms found within trees come from? Second, now ironically add, where did the carbon atoms in fossil fuels come from? Yep, the exact same place…the atmosphere. Via sequestration.

In all plant life and fossil fuels, something like 98.9% of carbon-12 atoms come from the atmosphere and nowhere else. They come from sequestering CO_2 from the atmosphere. Not from the ground the tree was planted in and not from incredibly lousy science saying otherwise. Fossil fuels are of course long-deceased plants or animals. So, that same CO_2 source made those plants and animals that eventually transformed into hydrocarbons as fossil fuel. After heat and pressure has been applied for millions of years that is. And yes, I am aware that hydrocarbons can form unrelated to carbon holding fossil deposits (AKA Abiogenic petroleum). Which is an exception and not the rule.

Moving on, trees get less than 3% of their total mass from the soil or ground they're grown within. Tree or plant mass (or their size and weight) all comes from sequestering CO_2 from the atmosphere with photosynthesis. H_2O (water) is extracted from the soil and part of the process, but CO_2 or the C-12 is not from soil. The H_2O a tree needs is transported via roots and also carries soluble nutrients up from the land. Those nutrients might have some C-12 but, as I mentioned, not much of the plant or tree's size and mass comes from the soil. Nutrients do help photosynthetic production grow the tree. So, they do help trees get bigger, but they don't supply a lot of the carbon it needs to form its structure.

Trees and plants also deposit enormous amounts of carbon into the soil in the form of growing their roots with atmospheric CO_2. Need some proof? I'm glad you asked! Observation of that fact is provided as growing tree roots lift sidewalks, driveways, roads, and yards as they increase their root mass. So, trees add root mass underground and move things out of the way as they grow from photosynthesis's use of CO_2. Roughly half of that mass deposited to move things is carbon-12. It's not magic, superstition, politics, nor internet influence. A growing tree or plant takes CO_2 directly from the atmosphere, so the plant or tree absorbs, uses, and grows from absorbing CO_2 from the atmosphere it interacts with and turns it into its carbon containing woody biomass, like roots, trunk, limbs, bark, and leaves or needles, seeds, cones, flowers, fruit, nuts, sap, etc.

Going one step further, trees and plants absorb CO_2, break it down, and store the carbon-12 atom, while releasing oxygen (O_2) for us to breathe. This mass can come from C-12, C-13, or C-14—carbon isotopes that photosynthesis can use or create through chemical reactions. There are fifteen carbon isotopes in total, but these three naturally occur and are utilized in making trees, plants, and fossil fuels. Interestingly, even carbon monoxide (CO), a flammable gas associated with car emissions, can be used in photosynthesis. Cool, right? All this makes photosynthesis the most effective tool in sequestration science and ending global warming!

The parts to remember is the tree does all this when the carbon atom was within the CO_2 molecule during photosynthesis and not because a carbon-12 atom has somehow miraculously appeared within its biomass or came from dirt the tree grew in or the water it used. In addition, and highly significant to sequestration and thus extremely relevant to CMS is the CO_2 molecule weighs 3.67 times more in amu when the plant or tree removes it from the atmosphere than the basic carbon C-12 atom the tree retains in its sturcture.

HERE'S A SHORTCUT

Read on for the summary or *skip to the bold* section below to get the gist. It's some empirical details about the 3.67 to 1; the CO_2 to C ratio and why it's significant in how it relates to global warming. It's about fundamentals that are ignored by many.

FYI, CO, like from a car or coal emissions, is 28 amu (atomic mass units, its weight of one molecule) and flammable! When it burns again it turns into CO_2 through oxidation. But we are going to stick with CO_2 because it is more difficult to deal with, obviously or we wouldn't have global warming. CO produces smog and is bad. However, it is less than 1 ppm in our atmosphere because it tends to bond with water vapor and become acidic rain, which is an entirely different problem than global warming. But it is a problem.

CO_2 has one atom of carbon-12 plus two atoms of **oxygen-16**. Simplified, $16 + 16 + 12 = 44$ amu. For precision's sake, 44.0095 g/mol. That is the mass of one molecule of CO_2. And I'm not picky, so please think of mass as weight, if you need to. Anyway, a metric tonne of CO_2 contains 22,730 moles of CO_2. To make this short, One CO_2 ppm in our atmosphere is equivalent to around 7.8 gigatonnes of CO_2. FYI, one metric "tonne," Is 2,204.6 U.S. pounds, Lbs. One gigatonne is then 2,204,600,000,000 Lbs. (2.2046 trillion, $2.204\ E^{12}$, U.S. Lbs.).

Sequestration's power starts to come into focus with the difference in amu between the single carbon-12 atom and the CO_2 molecule. CO_2 is 3.67 times more in amu, or mass "generically as weight," than carbon-12. As Yoda would say: "Chemistry 101, it is." Okay, maybe not. I'm compelled as a scientist to explain, weight is not mass but thinking of it like that works for this example. So please do. For the record, weight implies gravity, and mass is measured (weighed) without gravity. Here's the part CMS details as important.

CMS UNDERSTANDS THAT AMIDST ALL THE UNREASONABLE AND UNINTELLIGENT NONSENSE IN THE WORLD, ONE ABSOLUTE TRUTH STANDS FIRM NOW AND FOREVER: "WHEN A TREE ADDS 100 LBS. OF CARBON-12 ATOMS DURING NORMAL GROWTH, IT HAS ABSORBED AND TRANSFORMED 367 LBS. OF CO_2 MOLECULES FROM THE ATMOSPHERE TO DO SO." WHETHER IN POUNDS OR KILOS, THE 3.67:1 RATIO REMAINS CONSTANT. FUNDAMENTALS CAN INDEED CHANGE EVERYTHING-IF ONLY WE APPLY THEM.

For the record, I've seen the ratio depicted at 3.58:1 in atmospheric CO_2 to C examples. I'm not sure how that's possible but I've seen it used a lot, so it's worth mentioning. I mean 44 amu divided by 12 amu is 3.66666. Anyway, the result is always going to be empirical to photosynthesis's use of carbon dioxide at 3.67 or 3.58. It will always come out the same no matter how it's measured with one exception. The exception is excessive CO_2 fertilization in photosynthetic production which increases the ratio higher and is not a good thing. I address that ugly point in more detail later as an *accumulating decline*. Right now, I'd like to explain this part's importance. So, let's expand that 3.67 to 1 empirical nature to the study's bigger picture.

That 3.67:1 ratio expands easily from the micro to the macro. It scales up nicely from one leaf on a tree to a tree with thousands of leaves. That scaling also applies from that one tree into a forest of trees. In CMS's study, that ratio scales up from one forest into what's left of the Earth's forests. All because a growing tree adding 100 Lbs. of carbon atoms within its annual growth cycle is nothing special at all. I mean it is special, just not unusual. Some tree species will add hundreds and many tree species over a thousand pounds of carbon-12 each year they grow. That upward scaling ability is a fundamental, all because of something else that is absolutely needed for trees to scale it up.

Scaling up sequestration in a tree, a forest, or Earth's remaining forests requires "maturity." Maturity within a forest or tree regulates the amount of CO_2 absorbed from the atmosphere. Which also implies that "maturity" in any tree species is proportional to the tree's size (mass). Intuition and intelligence even tell all of us the older the tree, the bigger it is. In CMS's understanding, the older the tree becomes the bigger the tree <u>must</u> become. The fundamental in that is most don't understand the bigger a tree becomes the more it must grow during its next growth cycle. It's forced to do so, or it dies. Woah, that sounded a bit off after I wrote it. Maybe even counter intuitive to trees becoming bigger. I mean, they should want to grow, right? Yes and no, the inside of the tree is already dead but preserved by the outer growth. As in, the inside is not growing anymore, only the tree's outer portions grow wider, project limbs out, push into the soil, or reach for new heights.

To explain further, the tree's annual growth rings producing every year make the maturity relationship another fundamental that is often discounted or overlooked. You can count growth rings in a tree's stump and tell exactly how old a tree is or was. But if you count those rings halfway up the trees height it's not an accurate age record. Trees grow by adding a growth ring to an upward pointed conelike structure. So halfway-up the tree won't have all the rings the stump's true age record has.

The same goes for a trees limb age. You must count closest to the tree. Limbs grow just like trees but horizontally. Limbs on an old-growth tree can be entire forests all by themselves and why old-growth is important to planetary sequestration. Their canopies and limbs create a hanging forest and use very little ground area to generate massive sequestration levels. I call the old growth limbs Earth's hanging gardens of Eden.

As you count growth rings did you happen to notice those growth rings increase their diameter every year? Yep, the tree must get bigger every year, or it died. It preserves the non-growing portions as it grows bigger. It encapsulates them. The cone's base diameter and height increase to accommodate growing around the previous year's cone to form the ring. Think of it like stacking traffic cones, except the cones are larger each time you stack one. That doesn't account for new limbs, new heights (tops), or all the other things the tree grows with the carbon-12 extracted from CO_2, like roots. So, all the tree's parts increase proportionally with maturity.

Now for a regional shocker to people who think they can live with global warming by moving to the arctic regions that are warming. It won't work. The reason arctic regions don't grow big trees, or anything but shrubbery is because they don't get enough sunlight when temperatures do allow photosynthesis. Global warming temperatures don't produce more sunlight. Sunlight is not optional in tree growth, it's required. That is due to Earth's wobble and that (precession) won't be changing for a billion years. So even though global warming is melting artic ice and tundra it isn't growing more of anything big enough or useful in surviving global warming, like big CO_2 sequestering trees or the crops needed. Plus, the *accumulating decline* from the arctic melting is releasing enormous amounts of CO_2 and flammable CH_4. It will be a gas-filled swamp that is uninhabitable and too dark to even grow a carrot. Not to mention the lack of topsoil.

Sometimes growth alone is not enough to make the tree larger. Trees have other needs to perfect their growth rate and achieve their sequestration potential.

REGIONS AND SPECIES ROLE IN CMS.

Tree growth rates in an arid or arctic location could take three times as long to mature sequestration abilities compared to tree growth rates in a rain forest or coastal area. Literally, a tree at age 30 located in a dry desert or cold arctic region could be the same size as a 10-year-old tree in a rain forest or wet coastal range. Although their mass may be the same and they could both be sequestering the same amount of CO_2, the region they grew in regulates their ability to grow big and their growth rates to get bigger.

Unfortunately, the years needed to develop adequate sequestration levels can be immense in nonproductive regions that still grow trees, like high deserts on the west coast of the USA. Regions can therefore make tree maturity far more important in some places than others. Some regions are just faster at growing trees and sequestering CO_2 as a result. And yet, trees in unproductive regions may have the same CO_2 sequestering abilities as *sequestration value*, should they achieve maturity and old-growth status. Even nonproductive tree regions can have good enough environments for tree growth, although slower. Thats because they have the right species of tree for that region.

The gist is that harvesting trees in those arid and arctic regions is by far more impactful in forming climate changing conditions. Harvesting can also be considered far more costly to those not very productive regions due to regeneration time. But that doesn't stop us from harvesting them, not yet. So, slow tree growing regions should be managed with more time and, in many circumstances, not harvested at all. Even thinning under *natural attrition harvesting* that CMS promotes comes into question within slow regenerating areas. The time for *sequestration value* to arise is too great to mess around with any harvesting. However, sometimes resolving that issue comes down to the tree species available.

Accordingly, some tree species are also better than others at maturing within different regions. All tree species have limits to their height, mass, and age obtainable. Trees can also be restricted genetically to where and how well they will grow. It's in their genes. For example. I've seen acreages of tree stand failures because replanting took place at the wrong altitude, slope facing direction, or choice of species. Or all those issues combined. Those same trees replanted at an altitude of 1,000 feet less and on a north facing slope flourished. What limits trees is usually tied to the region and the species. If replanting is done, we must keep in mind it's not up to us what grows where when planning for maturity. We do need to recognize that more than ever. All because some trees are more marketable than others, we have constantly violated that forestry rule with *demand driven forestry* making all the wrong decisions. The impact? Doing so has spent maturity, biodiversity, and biomass production like a kid in a candy store is willing to spend their entire allowance. For the long term good of forestry regions, or even your front yard, always keep in mind the biodiversity of that region is just as important as the region that has trees at all.

Native species rule the requirement to create a healthy mature forest meant for centuries and not a few decades like today's forests have. To be fair, this is not a requirement to resurrect the Titans into the atmosphere regulating forest we need. At this point, we'll take any tree over none at all. However, replanting non-native species should be abolished in many regions for no other reason than the biological mess and long-term maturity setbacks they often create. Non-native impacts are difficult to reverse because there is often a financial reward that goes against them. That is *demand driven forestry* and severely limits *sequestration value.* The setbacks in maturity can be measured in centuries with the native species usually winning over nonnative replants anyways, so why do it? Economics, greed, you name it. However, *Natural attrition* is on the native species' side, but it could take centuries before its victory is ensured, so why fight it for a few coins now then more coins later? All this boils into the fact that some tree growing regions and species are just more sensitive than others and native species rule when it comes to enhancing maturity. But not always. There are exceptions I need to bring up.

FOR THE CMS RECORD, DON'T DO IT.

Am I saying stop replanting non-natives and cut down all the non-native species out there? No, and I really mean no! We'd have no forests left if we did that. Unfortunately, I've seen some organizations doing just that in Scotland and many other places citing wildlife enhancement as justification. Perhaps they'd change their mind if they understood just how lacking sequestration is and that every mature tree remaining counts endemically. Don't remove a tree, native or not, unless it is dead or dying and is within *natural attrition harvesting* criteria. Especially if it's 20-30 years or older or growing well. Instead, use *natural selection harvesting* and replant with the natives around the non-natives. That will allow natural attrition to run its course while generating biomass and maturity. All the while, it can unset the nonnatives with the native's genetic superiority. And I understand that it may not work in all scenarios. Some species are better than others at occupying ground. If so, it's better for all our sakes to let them mature or have replacements mature before dealing with them otherwise.

What I'm trying to say is we need to think about what is being replanted and always default to native species. If it's not native now, we need to stop and think maturity and sequestration before removing /replacing it. Upsetting the gene pool or decreasing sequestration further with unmitigated removal lacks foresight. Both lead to sequestration we need more than ever being lost. That's time we cannot afford.

For example. Have a look at Germany's very mature but non-native spruce trees. In this case humans aren't given a choice beyond the bad ones made and perpetuated since WWII. Bark beetles are killing their non-native spruce trees by the thousands. Google that. It's a shock I'll be adding to later, it is part of the *accumulating declines* being accelerated. Now some say good riddance to the nonnative spruces, me, I say it makes global warming worse, measurably worse in ppm.

ALL TREE SPECIES AND ALL REGIONS.

As I mentioned, all trees, at least most tree species, put on mass with age. Understanding that the percentage of mass added annually may decrease by species, region, and over their lifetime, but added mass is always based on their previous mass and prior to the tree adding new growth. It's an undeniable fact that all trees get larger with maturity even when the other factors are not addressed properly, or it died. Tree or forest CO_2 sequestration increases every year the tree and forest are alive. So, if humans allow the tree and forest to mature, the relationship expands and regulates the levels of CO_2 in Earth's atmosphere. And that happens without fail. It also happens with or without us. It's programmed into nature's plant growth formula. Isn't that interesting? I mean, the older a tree is allowed to become, the better it is for all of us and Earth. It even gets better in many ways; AND at this point in global warming, we have a lot more to gain than lose! Don't cut it down unless its dead.

BIOMASS EFFICIENCY. A TOUCHY SUBJECT, LIKE SHARING A ROOM WITH A HERD OF ANGRY ELEPHANTS.

I'm now going to bring up CMS's conclusions about **biomass efficiency.** Lack of biomass efficiency is a highly relatable subject to climate changing conditions forming. For thousands of years, *demand driven forestry* practices missed the biomass efficiency point entirely. Being highly impatient with forestry regrowth or regeneration in general brought about what CMS coined as *constrained deforestation.*

I've become more realistic about harvesting trees by two facts CMS points out. The **first** one is a new understanding of the term renewable when used in the context of existing forestry management. Renewable no longer has the credibility it once had, because of the second reason. **Second**, an improved description of trees provided by CMS. Coined *as a binary restricted resource*, CMS proves that trees are made up of two renewable natural resources in one green package, biomass, and *sequestration value*. It goes on to state that both of those renewable resources are required to continue Humanity's ongoing self-domestication efforts.

Those two CMS biomass efficiency facts also point out realistically that using forests for Humanity's endeavors will never go away. It can't, because forestry biomass is a well-established resource with net negative potentials, which is the only one available to humanity. IF it's used with maturity curbs that works well. So, I accept the harvesting but only with CMS's curbs. I may not like what they imply, but I do accept them. Both explained, it's time to expand the concept.

What CMS and no one should accept is how we represent both renewables priorities in the global order. Today, only for quick profits and not climate mitigation or biomass efficiency. This book explains later in detail why that is a threat to forestry and still a compounding problem that includes a promise to evict Humanity from all future decisions.

For now, I'll propose a reasonable definition about biomass efficiency because it does not appear in this book's glossary. In short, as trees get bigger you need fewer of them to do the same thing that needs many smaller ones. That is; to keep our self-domestication with items trees create we can use one big tree or tens of small trees.

For example, let's say one mature tree that was replaced and accordingly died of natural attrition is thinned from the forest. That mature tree could have more or the same volume of biomass as 140 immature trees normally harvested today. If that isn't enough to make a point about lack of efficiency, let me add those 140 immature trees that required one acre (2.47 acres per hectare) of very flammable land to grow. That one mature tree thinned out by natural attrition occupied around 480 square feet (44.59 m^2) of sun and moisture filled land that was fire resistant. Mature growth plots are like a sponge for keeping water and the bark on older trees insulates them from fire. Natural attrition within a mature plot also produces more biomass in shorter durations than immature plots. But only if managed by natural attrition!

For example: There are 43,560 feet^2 or 4,046 meters^2 in one acre of land (2.47 acres in a hectare). So, in this scenario, there is enough room for roughly 90 super mature trees per acre. The point. Just one acre, with 90 mature trees is enough to do the same thing as up to 12,600 of today's regularly harvested, tiny, and immature trees using 90 <u>acres</u> of land. Maturity in this scenario is 90 times more efficient.

I must ask, are immature trees an efficient use of Earth's forestry resources? I don't think so and now know better. All because that one mature acre is regenerating biomass as the old ones and the stands more recently regenerated die off. Anyway, the plot ends up generating way more biomass than the 90 acres of immature trees can, before their harvested young, as small logs. All one needs to do is apply patience. Then thin the dead and dying out of a mature stand and we might never run out of biomass regardless of the population's demand for it! Isn't that something? Maturity ensures future forestry demand.

CMS's math also says heck no to today's forest efficiency and checks out as well. This becomes a little more or a little less given region and species. The downside is the time required for today's immature stands to gain enough maturity. Impatience can again reveal it's ugly nature and make it easy to ignore maturity, just as it does today.

THINNING'S ARE GOOD FOR EVERYONE AND NOT TREE PRODUCERS, YET.

Proper thinning's are the intended results of CMS stewardship models. They are only successfully achieved when the natural attrition rate of a forest is allowed to increase maturity and *sequestration value*. Applied correctly, thinning tree stands greatly improves CO_2 sequestration ability by allowing more limbs to grow. More limbs mean healthier, faster growing trees that are sequestering hundreds of percentiles more CO_2. But immature tree thinning costs more than clear cutting and initially produces a lot less usable biomass; but not always. If a mature timber stand is thinned profit does ensue as remaining trees result in increased biomass and production with increased sequestration. Therefore, even the first step in gaining forest maturity can provide profit. Plus, that modest efficiency gain beats out clear cuttings profit even if the stand is only slightly mature.

Knowing tree stands are immature, it is no surprise that timber companies don't do thinning's. They can't because there is no profit like the one clearcuts produce because maturity has been ignored. To CMS, thinning forests must generate revenue for the forest owners like clearcuts do now and without sequestration income they can't. But with, well, that future should ascend *sequestration value* and thinning's globally.

To be sure, today's forestry management is all quantity not quality. That is a direct result of not having mature trees and zero biomass efficiency in any forestry planning. This catastrophic form of forestry management is so engrained that it's now the only way the global forest industry believes they can survive and continue to thrive. Literally, quick profit is given priority when decisions about efficiency, climate, quality, and sustainability arise. The funny thing is, improving biomass efficiency greatly improves profitability. Seriously, CMS could make legacy forest land the most valuable properties anywhere, even more valuable than skyscrapers. So, the industry may be wrong today, but they will come around once they can see the benefits and profits. Plus, it's scalable. So, it does not have to be implemented all at once or with great sacrifice. And that's the nicest way I can say that without adding please.

Yes, the wood industry is a long way from improving biomass efficiency, getting richer, and ending global warming. Which is what CMS mitigation is offering the wood industry. The sequestration derived plans take time to convince others of the need. And the time needed to convince others? That is time we don't have much left of. Intervention is now needed, or else.

And this is where sequestration science takes off like a rocket away from the contemporary knowledge of climate change. It's also where my big ogrelike feet start crunching the toes of perception. Yeah, it's stuff that's difficult to believe… at first. It also ruins the forest in the postcard and the forests you like to visit. Again, my apologies for going rogue but as you soon or perhaps already understand, I had no other choice.

ROCKET TO ANOTHER DIMENSION OF THOUGHT.

Atmospheric CO_2 outflow

Currently, many science disciplines believe photosynthesis creates an atmospheric CO_2 outflow in a fixed amount and that fixed amount happens yearly. That outflow occurs during the annual plant growth cycles. It's currently believed to be as such because it's thought all plant life is too predominant on Earth to allow any noticeable fluctuation. Many scientists assume removing one plant allowed another to grow. Thus, the CO_2 outflow from atmosphere should be a constant of sorts. We really believed the outflow volume would not fluctuate, much if any at all. That is an easy assumption to make given what we knew, at the time. So, it is an assumption. An assumption that led to emission reductions being the only way to fix global warming. An assumption that says trees can't fix global warming. That assumption states there are far too many plants for a CO_2 outflow fluctuation to happen or at least a measurable one. And so, we believed wholeheartedly that the outflow of CO_2 in climate experiments and predictions would always be the same result achieved in annual plant growth cycles. Again, static, a constant, and fixed amount. But guess what, none of that accounted for the measurable amount forest maturity brings into that conversation. And it is very measurable, without assuming anything.

CMS proved otherwise because of tree growth's proportionality to maturity being proportional to sequestration ability. Times that by three trillion trees and saying we were wrong about atmospheric CO_2 outflow really isn't enough to forgive that still ongoing scientific omission. An oversight emission science still perpetuates. CMS's CO_2 ppm Deltas model explained the maturity connection to atmosphere. But that wasn't the only way.

It turns out, Earth breathes CO_2 according to the maturity of forests. So figuratively, Earth breathes CO_2 with forests acting as Earth's enormous lungs. Those CO_2 breathing lungs regulate Earth's weather, temperature, and most importantly rainfall. They are the one and only animal life support system on Earth at any scale. So, here's the problem we need to cure global warming. Earth has pneumonia. Earth's lungs are clogged with immature trees that don't contribute to removing CO_2 as they should. Even our mature trees are in a *carbon hump*. Add that human forestry demand gets in the way of resurrecting Titans, and we've earned the global warming trophy.

Today…better logic has built sequestration science's models and earned undeniable precedence. That wasn't so easy to do. Or to believe, at first. But the logic says it all and does so strongly enough to encourage certainty with the facts.

THE LOGIC OF IT ALL, TRUTH DELIVERED FACTS.

At one point during the study, sequestration's effect on climate change still needed words that didn't exist. At the time, its impact was observable and measurable, but I could not explain in words how that impact came to be or existed. Essentially, I lacked the logic that could explain the numbers. The logic of it all. That was something that had definitely not been defined by anyone, to include me. So, nothing about how forestry, maturity, current forestry management, or historic forestry correlated with atmospheric CO_2 ppm combined into explaining global warming. Really, nothing about how humans created climate changing conditions as an environmental impact. Unfortunately, all the reports available were either emission tilted or blamed emission's influence. They only counted-up emissions and referenced emission sources as the problem. I couldn't find anything that spoke of sequestrations grander impact that maturity most definitely provided. To be sure, there is empirical data on sequestration science, but it all seemed to downplay or not mention tree maturity, scaled or not. There is also empirical data on CO_2 ppm impacts and a forestry paper that gave the *carbon hump* more credibility, but not openly. That came as the paper was explained by its Author, Dr Law (1). So even the study's I found useful were emissions focused. But those emission-based study's scratching at the surface helped reveal sequestration's importance with empirical assessments of very specific topics, they helped a lot.

Eventually, I formed my own way of describing CMS logic. I had to. Humanity was basically missing a fundamental in describing climate change because it's not the climate changing due to emissions all those papers imply or say outright. It is humans using forestry to create climate changing conditions and that created global warming.

That was the first part of the CMS's logic to reveal, changing the way I thought about climate changes origin. That thought led to the remaining logic and proof. It was hidden in plain sight, and it wasn't in the last place I looked, I had to look everywhere and in places I didn't want to.

EUREKA! AND DARN-IT CAN OCCUPY THE SAME MOMENT.

The breakthrough came from an unexpected yet golden data source. I never anticipated that historical forestry data would correlate so well with atmospheric CO_2 levels. Forestry's impact on sequestration had been largely unnoticed or disregarded, so I dug deeper. Months of mining published papers revealed a shocking history of forest abuse. Historical reports and media often incrementally mention how 30% of global forests and 46% of trees have disappeared. You can even find dates to track these losses and estimate the age and harvesting date of the trees. And sadly, see the impact in atmospheric ppm within the same timeline.

For instance, Roman metallurgy and the Germanic wars are well-documented. Another, Pliny the Elder's writings, also a Roman is a reliable source. He detailed the Roman Empire's forestry practices and their brutality to acquire forests. Only to decimate them immediately after acquisition. Earlier still, the demise of the Lebanese Cedars tells a similar tale. Many of the globes empires expired and cultures disappeared to their deforestation practices. Most because they needed wood for metallurgy and massive construction projects; only to recognize the Lorax's warning too late. And history is still repeating today as too late for humanity now exams a more severe consequence. All because none of those forests mentioned in our history have yet to recover their maturity. Have we learned nothing, until now I hope?

China, first deforested to build enormous ships in the 1300's. Practices they haven't abandoned with their more recent economic booms. Europe was out of mature forests by the 1500s, with English, European, and American ship building and reliance on wood as fuel depleting global forests later. Eventually, we shifted to steel ships due to timber shortages. Firewood fueled Europe, they ran low in the 1600's creating the first energy crisis.

Spain's King Philip II used every immature tree available to build the Spanish Armada, resulting in substandard ships from already immature forests that burned in the English Channel in 1588. Spanish, Portuguese, and French forests still haven't recovered from his decisions. Demand for timber continued to outpace forest regrowth, particularly with the advent of railroads, which utilized all remaining accessible forests. Even modern records, like the last forty years of cypress logging in the lower Mississippi or the last old-growth logging on the U.S. west coast, tell the same story of *demand driven forestry*.

A story we can read about because conquerors, timber barons, and government have always boasted about their forestry conquests, repeating the same tale of deforestation for thousands of years.

CMS defines these events logically, using terms like *unconstrained and constrained deforestation* along with *demand-driven forestry*. Understanding this logic, supported by the book's glossary, reveals how historical forest demand correlates with atmospheric CO_2 increases. NOAA's (Pronounced "Noah", National Oceanographic and Atmospheric Administration) ice cores and Mauna Loa, Hawaii Observatory's flask method recordings confirm these findings.

Let there be no doubt that forestry history laid the tracks for global warming and accelerating our current climate issues. Atmospheric CO_2 levels and forestry decline are interwoven, with tree maturity as the thread. But this "quilt" isn't pretty; it's a stark and unsettling reality. Uncovering this correlation was shocking. It wasn't a coincidence; the numbers made sense and logically defined how these correlations became increasingly accurate. Every page of our well documented forest history guarantees it.

I was frightened at first by understanding the true cause of global warming. **First**, because it wasn't what we all believed. **Second**, this perplexing message made communicating sequestration science challenging against the prevailing emissions-focused beliefs. It's disheartening to see blank faces when explaining CMS and feeling like an outsider. Yet, pondering the next questions made the realization even more daunting.

CLIMATE'S ALREADY MURKY PERCEPTION

CLIMATE CHANGE IS NOT WHAT OUR PERCEPTION BELIEVES.

I understand, you probably think human emissions are causing the climate to change, so reducing emissions will fix it. But what we all perceive as "climate change" was actually geo-engineered, unknowingly, over thousands of years by humans modifying forestry resources globally. Yes, loss of tree maturity caused global warming. That describes climate changing conditions and not a climate change result. To explain, the climate is not changing willingly via emissions alone. Humans' are changing and have changed it. Now we've received the global warming trophy for our efforts. So global warming is a result of humans creating climate changing conditions by decreasing forest maturity. Now that we know exactly what we did for our trophy; climate changes arbitrary role in definition is over because global warming has been defined, correctly.

WHAT DID NOT CAUSE CLIMATE CHANGE?

The Industrial Revolution 1760-1840 CE was not when climate change started. It accelerated an already ongoing climate changing condition with a mechanical industrialization of forestry and transportation improvements impeding sequestration faster. But, as I stated earlier, global warming actually started thousands of years before the Industrial Revolution.

The industrial revolution began in 1760, 264 years ago. Atmospheric CO_2 ppm levels have increased rapidly since. Now think back to what I and the United Nations said about

30% of global forestry land and 46% of trees disappeared within the last 250 years. Hmm, a correlation, yes, a significant correlation. So, that rapid ppm rise wasn't all due to human emissions levels increasing. The rise was caused by the growing forestry demand cutting mature trees even faster. In that timeline, *Constrained deforestation* was being spread globally. Therefore, during that time, human's began harvesting global forests quicker and at steam powered industrialized scales. Add that wood was the fuel of choice for most of that timeline as well, while keeping in mind coal or fossil fuels, didn't replace wood at scale until the 1890's. That's right, 130 years after the Industrial Revolution began. However, it's fair to say that revolution hasn't stopped. Its only switched to different countries and different fuels. Ahh, coal use, right?

    The main reason cited to avoid the coal transition in steam powered railroads, ships, and factories was, "wood comes from everywhere and coal comes from a hole in the ground a longways away." Subsequently, the fossil fuel transition didn't happen until *convenient forestry* literally ran out of wood and could no longer supply the steam driven growth orientated economies. In short, coal was harder to get, more expensive, and not as good as a fuel as wood is. So, they used up every *convenient forest* to power the steam powered revolution, quickly and without foresight. Even some of the first electrical power plants used wood as fuel to make steam. But the industrialized nations ran out of wood quickly.

    It's easy to misinterpret the Industrial Revolutions data using emissions but the Industrial Revolution wasn't when climate changing conditions or global warming began. One must look closer and be aware of sequestration's downfall to see the bigger picture.

    And I can't say this enough, emissions are a problem, but not the entirety of it. Emissions are an input, but they lack precedence in actually causing global warming. Just as much as the period known as the Industrial Revolution did. Yes, emissions increased by magnitudes, but sequestration had decreased my millions of percents by then with much of the world's forests becoming super immature and kept that way.

So, the Industrial Revolution changed sequestration more than its emissions affected the atmosphere. The first industrialized and steam powered sawmills could chew through trees faster and those trees could be harvested and transported from far away; globally, *inconvenient forestry* was becoming *convenient forestry* as advancements in transportation developed. Shiploads of wood are historically recorded as they left the America's every day to go to Europe. Much like todays shipments of grains and wood to China and Europe. Proof is the correlation between forestry demand and atmospheric CO_2 ppm increasing in a unison. That union developed over thousands of years prior to the Industrial Revolution but accelerated then as it still does today.

ISOTOPES.

How do we know emissions increasing in the Industrial Revolution and other times isn't the cause? Good question. Here is another way to answer but there is another way that's even better. We literally have multiple proofs. However…

Differentiating between historic ppm levels and emissions gets tricky but solvable. Fossil fuel use helps to tell the difference by analyzing carbon isotopes. Fossilized hydrocarbon emissions leave a fingerprint in the atmosphere and within the ice core samples. What fossil fuel doesn't have is the ^{14}C isotope, Carbon 14. You might recognize Carbon 14 as the way to find when an item was created. Using Carbon 14 dating. To keep this simple and away from ^{14}C's basic ability to date back 62,000 years and advanced ability to go back even more, fossil fuels are millions of years old so the radioactive ^{14}C isotope doesn't exist in fossil fuel emissions; it decayed over time into ^{14}N, Nitrogen 14. As a younger organic matter 62,000 years of wood emissions have ^{14}C. We also see ^{13}C in our atmosphere. Volcanos emit a lot of ^{13}C. Anyway, we can see the difference in isotopes within ice cores samples historically and today's atmospheric samples.

Not differentiating isotopes and seeing the CO_2's atmospheric build-ups led science into thinking fossil fuel emissions are the cause. As I mentioned, an easy misinterpretation to make when you don't consider sequestrations impact or photosynthesis's ability to recycle ^{14}C into just plain ole C. Yes, it does that.

Thousands of years ago, emissions from wood fuel were increasing as populations grew. However, there weren't enough people on the Earth burning wood or making charcoal to make an emissions difference. And yet, the atmospheric CO_2 ppm was steadily increasing. That can only be answered by sequestrations decline. Add fossil fuel. It's use started gradually in the 1600's but didn't take off until the 1890's and you get the same conclusion, sequestrations decline. To be sure, emissions were increasing as the human population grew and fossil fuels were being adopted but sequestration decreasing makes a more significant impact on global warming forming and maintaining by increasing atmospheric CO_2 concentrations, as an isotope or not.

Yes, there is no denying that CO_2 emissions are accumulating in the Earth's atmosphere and causing global warming. But that alone does not paint the big picture that sequestration's decline does. CMS's proves many points to help judge that argument. One of which states,

CLIMATE CHANGE BEGAN WITH HUMAN SELF-DOMESTICATION EFFORTS THOUSANDS OF YEARS AGO BECAUSE GLOBAL WARMING IS NOT SOLELY DEPENDENT ON CO_2 EMISSIONS. CLIMATE CHANGING CONDITIONS ARE RESPONSIBLE FOR THE DECLINE OF CO_2 SEQUESTRATION ABILITIES (AS FAST-CYCLE CO_2 TERRESTRIAL SINKS, MOSTLY FOUND IN FORESTRY).

Therefore, the beginnings of climate changing conditions and global warming are summarized as both *sequestration and emissions dependent*. Which is a term CMS coined to help describe it's logic. The two must be in balance for even modest atmospheric homeostasis. However, because CO_2 sequestration relies on atmospheric residence durations and forest maturity, sequestration ability is more consequential in creating the climate-changed conditions impact, global warming. In fact, climate changing conditions are more responsible for global warming than after combining human and natural CO_2 emissions are. I also provide another reason later. The isotopes present, well that is made entirely moot.

Note: Natural emissions are highly related to plant growth cycles like leaves dying, crops or tree harvesting, and forest fires. Animal quantities on Earth also play an influential role through respiration and food energy. In addition, geological releases like volcanos, ocean methane, and geysers are all sources of natural emissions. If man can't directly influence it or it is a part of life existing, call it a natural emission.

FOR THE RECORD, THE RESPIRATION OF ANIMALS ON EARTH CREATES AN ESTIMATED 7.3 GIGATONNES OF NATURAL CO_2 EMISSIONS ANNUALLY.

Human emissions are human-made and unnatural to Earth. In no particular order. They are mostly tied to fossil fuel production and use, construction, energy production, industrialization, manufacturing, and transportation. A large human emission source comes from forestry harvesting and wood products manufacturing.

CMS considers both natural and human emissions as unavoidable. To try and define one as worse than the other becomes moot in the sequestration picture, and why we still have global warming. That is meant to imply that any mitigation has to directly impact both emission sources and one other thing. The accumulated ppm level within Earth's atmosphere, 1,541 gigatonnes CO_2 in residence conditions. There are over 3,300 gigatonnes total but only 1,541 gigatonnes need to be dealt with today. Only 1,541 gigatonnes? Okay, I agree, that's a lot of CO_2.

Let's face CO_2's facts together. To be successful at mitigating, fixing, or curing global warming all three CO_2 sources must be addressed. And human emission reductions can only affect human emissions and nothing else. So, natural emissions and the even larger atmospheric accumulation are not impactable by human emission reductions. Not by any means, not even in the wishes or the dreams of emission reduction science is it possible. More on this is coming. For now, a more recent piece of logic to ponder.

CO_2 EMISSIONS ACCUMULATE IN OUR ATMOSPHERE BECAUSE THEY HAVE NO OTHER PLACE TO GO, BUT THEY USED TO AND CAN AGAIN.

That is far from today's thinking, but sequestration facts and logic ring that truth bell loudly. That truth is muting emission reduction's repetitive and offkey song and we should continue to let sequestration ring.

CMS DOESN'T CARE WHERE THE CO_2 CAME FROM OR HOW MUCH THERE IS. CMS CARES ABOUT WHERE SEQUESTRATION SCIENCE CAN PUT ALL OF IT WITH FOREST MATURITY.

That says a lot about global warmings origin story without saying it, doesn't it? Sequestration science is not about directly reducing emissions, because the climate changing conditions that formed global warming don't care about emission levels. Forestry sequestration sinks evolved to absorb way more CO_2 then any emission combined sources can produce. Which is why sequestration is the only way to remove CO_2 from our atmosphere while emission sources continue to add to it. And again, human emissions are bad and still need to go away. But they have no effect on fixing Earth's sequestration problem, nor the global warming impact sequestration's downfall created. Resurrecting Titan's is the only way to fix it all.

WILL WE USE IT? THAT QUESTION INVOKES ABSOLUTE TERROR.

The math worked, the correlation evidence was in place, and I defined the logic, and I am ongoing with the CMS study. There is much more to do. What I then ended up doing was coining terms to explain sequestration through CMS. The logic within those terms is also found within this books glossary. I hoped to explain CMS enough to fuel constructive arguments. To explain, I'll define another Eureka moment days later, as it blindsided me.

CMS is the first to describe exactly what caused climate change and how to fix it permanently. Or did it? I'm certain about the undeniable results. I mean, trees are defined empirically and what CMS says about them is therefore pretty basic. It's all fundamentals assembled by a logical order of operations. Now, the atmospheric science in CMS is not so empirical and is ongoing work in future climate prediction. But even that is accurate and reproducible. Those reproducible results say a lot of good things about CMS credibility.

Combining it all with logic and proofs, still seems basic math and science but putting it all together, well, you be the judge. I know CMS and I were the first to state climate change by definition and reproducible exposition. CMS findings date back to 2018, I wrote a related hypothesis in the early 1990's. But honestly, I really don't know, and I really don't care. I just want people to use it. In these times of academic overcrowding and studies of study's, who knows. What I am certain of is the logic and proof say a lot more than we ever wanted to know. Plus, they're going to keep talking whether we agree or whether we like it or not. Fact is like that; it sticks to all known surfaces, like it or not and eventually it will stick.

So maybe a new scientific discipline? I can't be the judge of that, I'm biased. But I did find it, so I guess I could also name it. I began to refer to my work as Full Mitigation Science or FMS for short. That got your attention, didn't it? Everyone balked at that name and said, "Complete Mitigation Science," or CMS for short is more descriptive of its results. It does seem to fit, and better, so I adopted it. And with the name, dreams of the future began to appear. A future without global warming, a grandiose recognition, and wide scaled acceptance of CMS facts with gratuity. Finally, a justification for my years of CMS induced struggles came alive in my dreams. Yeah, right, I'm not waiting for any of that, time is too precious, and I know better. Still…potential arises, and dreams do form from even less.

But it isn't over until it's over- *Yogi Berra*

Eureka again! Another manuscript was born. This one is concise, "Sequestration Computation." It aligned my work with basic math and made it easier to view sequestration's impact and how to add sequestration into formulaic climate / atmospheric equations. But it also tells how much trouble we are really in. So, my dreams of soft furry forest creatures, flowers, and butterflies were crashed by a combined assault from emission reduction science's perception of climate change and now a doom and gloom CMS sequestration computation foretells. And with sequestration computation the nightmare of the will we use it question gained realities grimace. The abyss enlarged and darkened for two more years as I explained CMS to others before going public. All so I, and others could check my work. But now it's undeniable and we are running out of time.

Okay, maybe none of that was very modest, but how else can I say it? To justify that and the coming arrogance I need to explain one of the simple proven facts sequestration sciences revealed to CMS and now, to you.

UNMATCHED POWER OF SEQUESTRATION

Imagine combining all contemporary CO_2 emission reduction schemes from global climate mitigation efforts. Even without incorporating CMS, sequestration still outperforms all those combined efforts by a landslide. All of them totaled together! Sequestration science is undeniably the most powerful global warming mitigation tool ever proven.

Sequestration math is so straightforward that children can demonstrate its power. Google Maps reveals the truth about forest maturity—there's no hiding from it now. There's no other way to halt the climate-changing conditions causing global warming. Yet, we continue to march to the tune of emissions, heading straight into an abyss.

Today, the number of Titans is shrinking. *Accumulating declines* growing. Forests disappearing. Tree's harvested younger. The weather is becoming increasingly extreme, with hurricanes breaking records and causing widespread devastation. As I write this on October 11, 2024, twenty-four million acres of the Amazonian Forest are burning-an exceedingly rare and catastrophic event. These wildfires will rage until the fall rain arrives in months, if the rain comes. *Accumulating declines* have been given free rein, promising even more chaos next year.

Only sequestration can stop this, but we continue to ignore and question its effectiveness, we will do that until it's too late. Well, our empathetic exercise in existence is no longer tolerable. It's time to rudely wake everyone up! All because the next sections answer to the Lorax.

SO, WILL HUMANS USE SEQUESTRATION SCIENCE?

THE CURSE OF SEEING THE FUTURE

Figurately, I suffer from the same curse Apollo placed on Cassandra. With sequestration science's help, I can foresee the climate future, but I'm cursed so nobody listens until it's too late; and that becomes their regret-not mine. That curse creates my nightmare, one I hope doesn't become our reality.

My true curse? My Grandfather once told me I'd suffer the curse of being reasonable and intelligent, unlike my more popular and better-looking relatives who could get away with murder. They, he explained, relied on vanity to prove their competence, and make their way through life. "They have the advantage," he said. He then apologized for passing down this curse, saying, "It's definitely not a gift, Timmy. You need to be careful when you show it." He was right. So, here I am, a rogue scientist. To most academics, I'm just an engineer-a specialized simpleton easily ignored. I don't distinguish myself until absolutely necessary to defend facts. It's a defensive strategy, usually ruined when my competence displaces other's opinions with facts. Yeah, I'm the guy that doesn't suffer fools or vanity displays. Because I'm not ruled by emotions or peer pressure.

Usually, I present my findings and then watch as others predictably crash and burn by making avoidable mistakes. I learned early to avoid problems from ridicule and the incompetent expert ruining things that worked just fine. They always fail by ignoring history and science. I learned early to capitalize on those who ignore my and their own grandfather's curse, because their charred remains held profit.

At first, I adjust my behavior and communication to give others time to absorb my insights. Still, my advice is often misunderstood or unappreciated; until it's too late. Rarely do I get feedback on complex issues like this one, but when I do, it makes me happy knowing someone listened and has engaged. Even if their argument is baseless, their interest is usually not. That gives me hope in a change to come, a foundation to build upon. I live for those interactions, and I'm thankful for them. Although, I more often watch the predicted crash, I do from a safe distance, I've always tried to plan ahead. So, I'm usually the first to benefit from the crashes aftermath.

This strategy has been profitable but limited in gaining friends or supporters. Most relationships decline as the predicted explosion recedes; those who ignored my advice are too embarrassed to reconnect immediately. Some call later, needing a required expert on the team, not really seeking solutions just a checkmark in the Engineer box. So, they often ignore my advice again and again. But I still show-up when invited; knowing they haven't changed but hoping they do. The cycle of ignoring good advice for short-term gains defines stupidity, which is hurting themselves and others without any profit. But its proven to benefit me later, so please don't listen. Their minds were made-up before hearing me and that was fine then just not anymore.

I wish to escape from that short-sighted world because I know we can be so much better. I've seen it a lot, and it's wonderful. Not all my interactions required my buzzard like secondary planning. Now is not the time either.

Here's the problem with the sequestration science news I must share; I can't use my usual sit-back strategy to fix global warming. The carnage from a climate disaster has no profit to dine on later. We're all joined at the hip to this crisis; when it explodes, my family and yours will become casualties. There is no escape, no safety net, no severance agreement; everything will be gone. There won't be any bones left to pick over, nor any buzzard to do the picking.

For the first time in my life, I've become desperate in having my grandfather's curse. All I can do is present the facts and hope the world acts. From this book forward, it's out of my control and without a buzzard like fallback to implement. So, I'll serve up the cure and warnings as best I can and wait for help to arrive. And I do need help.

History suggests that CMS will only be considered after desperation takes hold; when it's too late. So, all of you cursed like I am, take notes and get busy. I know you're out there, hiding like I did. I've worked with and seen many of you working, fighting the good fight, and winning. We can no longer hide our curse. Our curse needs to be heard, and loudly enough to engrain sequestration into global agendas. Will we use it? Probably not in time. This is the Greek tragedy we face- the tale of failing to cure climate change quickly enough. Except no ones left to tell the tale.

THE SEQUESTRATION SIDE OF EARTH'S CLIMATE PARABOLA IS QUICKLY CLOSING IN ON ZERO WHERE THE CO_2 EMISSIONS AND ATMOSPHERE RESIDENCE SIDE CONTINUES TO STRETCH ITS VERTEX TOWARDS INFINITY.

In other words, it takes decades to implement sequestration curbs. So, CMS lacks instant gratification like solar panels tell of saving money or trendy electric cars that look shiny and zoom. Nor does CMS hand out anything for free. CMS is the solution Earth needs, but it's not the solution people want. So, CMS is tough to communicate, spread, and implement. To be sure, nobody wants sequestration's impact to be true, but it for darn sure is and it is downright scary! Facts and truth usually are, but nobody wants to deal with the reality of their ID being told it was wrong, and then forced to accept it. Yeah, thats going to be a popular adjustment to the globalized climate ego, not. But it's just the facts, so we'd better get used to it, and fast.

LET'S TALK ABOUT CLIMATE DENIERS. PAST TENSE.

At least some of them. The smart ones with Grandfather's curse. Here's something you don't hear often from a tree hugger: they were absolutely correct about emission reductions. Yes, you read that right. Climate change, as presented today, is incomplete and presumptive, until CMS. So, I'll extend an apology on behalf of science (as if I could) and commend these deniers for their instinct against the incomplete science being emotionally pushed on us (because they didn't know the complete picture). Deniers were right about emissions science and reductions being inadequate.

How can you blame them for applying common sense, even if it seemed like a gamble with all of our futures? They were intuitive about the real climate situation-they knew it was wrong, then. I can't and won't fault them for that. So, to the deniers, I say, "We missed you while the rest of us made the sacrifices needed to get to the next battle." Your absence has been felt, and your insight is now needed even more. The truth of the matter has been made clear. Because the fifth part is true.

THE FIFTH PART'S TRUTH.

Allow me to explain myself; if all the other parts of a five-part emission-based study are not creditable, but regularly communicated as such, then how can you believe the remaining part is? *The fifth part is true*. Because that fifth part of the study wasn't the presenter's findings, it came from multiple credible sources. It was used to sell the not so credible sources lines. So yeah, I get the deniers abandonment from a fight based on one credible fact but then related to many half-truths and gaslighting. But it is time for all of us to regroup and conquer, or else. That fifth part was the only truth in the emissions reduction statement. It's the part that is empirical in measurement and accordingly a fact. It is the rising CO_2 ppm level. Unfortunately, that precedence also spawned four embellishments. The embellishments deniers justifiably determined as wrong. Perhaps, and maybe, only communicated for some profit scheme or another is what tipped them off. Even so, that fifth part doesn't go away even if the other parts proved inconclusive, as assumptions, or as a bull fertilizing the pasture.

Here's some of the denier's facts that went to bed without supper. First, they knew CO_2 emissions are unavoidable. Yes, we are an *emissions dependent* species. Another fact they stated, you can't make accurate climate predictions, so climate change doesn't exist. True, when solely based on emissions you can't. We prove that in a bit with sequestration added in. Another fact deniers stated, emission reductions don't do anything. And yes, we'll show this truth as well, but they do have some impact, just not very much. And the big one, eliminating fossil fuels is only a paradigm to shift political control. Yeah, I'm not political but yes, that in CMS hindsight seems true as well. Perhaps because they didn't understand sequestration when those decisions were made. And now that fifth part. That part still stands taller by moment, its bankable.

The fifth parts truth, Increasing CO_2 in our atmosphere is problematic and will cause our extinction. That fact and fifth part's entirety is what it is and just as truly undeniable as CMS's ability to fix it. The difference in science method that regroups our army is CMS proved everything it relates, even to earlier deniers who proved all of us emission reducers wrong, which includes my own admission to that defeat, in exchange for admission to sequestrations sciences rise. I apologize to the climate change deniers (for what that's worth); you were right and you're not just another brain-dead zombie after all. Well, at least most of you aren't. That is, if you understand that the fifth part and sequestration science is as real as it gets.

THE FIFTH PART'S BLANKET AND DENIERS ARGUMENT.

For the record, the CO_2 volume of our atmosphere is 0.044% at the time of writing, at 427 CO_2 ppm.

I need the blanket you slept under last night to explain this. If you look at your blanket in comparison to the room and the house you slept in, that blanket took up hardly any cubic volume of either. As a percentage of the homes volume, let's say it's 0.002% of a decent sized home. Of course, the home internal volume influences your bedroom's and your personal temperature. The larger the volume of the home, the more heat is required to warm it.

Let's think of ourselves as being Earth in a dark and freezing universe. Well, that blanket did an excellent job of keeping us warm while we slept and while thinking about ourselves as Earth. Now, let's just be ourselves waking-up on a cold winter morning and remembering we didn't pay the gas bill. Even if the rest of the house and the stuff in your room froze from outer space's ridiculously freezing temperatures, you didn't. You didn't because atmospheric CO_2 is Earth's and your nighttime blanket. It doesn't take up much volume, but it retains the heat generated from the day to keep us from freezing at night. Night of course is when we only have the blanket to ward off the freezing of space that is dead set on removing all of Earth's solar accumulated warmth. CO_2 does a surprisingly excellent job at keeping Earth's warmth until the next sunrise. Just like clouds that make winter nights warmer, the CO_2 blanket does the same. Except, CO_2 is translucent, unless it is frozen. So, you can't see it doing its job. But you can be assured it's there, or plants wouldn't exist.

Compared to your volume in that unheated freezing house's atmosphere the CO_2 blanket's volume is hardly noticeable, but your Earth like warmth is. Without that blanket you would have frozen. Let's think of that blanket as 230 ppm of CO_2. That's how much is needed to cover you. So, at today's 427 ppm CO_2 and 47.75 ppm CH_4 to CO_2 equivalent. It is really 474.75 ppm of two greenhouse gases. Which is 0.056 % of Earth's atmospheric volume. That is not a blanket anymore it's two blankets plus and way too hot to sleep under. So, we need to get rid of a blanket and wipe the sweat away before going back to sleep, yes?

Our blanket also comes with a warning label. It says, "global average temperature is inversely square to CO_2 ppm levels and not directly proportional." That's important because it is a positive feedback loop; warmth creates more warmth which increases temperature much faster than atmospheric CO_2 ppm increases.

With temperature related here's the or else. When, or now, if, CO_2 ppm hits 840 ppm *accumulating decline* teeters at unstoppable. If it hits 940 ppm, it is then entirely unstoppable. I will explain more later. For now, if our 0.044% of atmosphere volume, our blanket, continues to get thicker, you'll be roasted at 350° F, 176.67° C at 1% of atmosphere. If you somehow manage to live through that oven setting, at 7% you suffocate from hypercapnia. The scary thing is if we hit 1% were only months from 2% and then weeks from 3% and then days from 4% and so on. That describes a greenhouse gas runaway. It won't stop until it reaches a saturation level, and it accelerates constantly to get there.

The point, somewhere around four blankets, human extinction is no longer a threat, that unfortunately becomes imminent with no way to turn it around, there is no foreseeable way back, period. In addition, medical study's tell us just before 1,000 ppm of CO_2 and oxygen breathers experience nausea and severe headaches. After 1,000 ppm, they die. Need I remind we are almost halfway there already?

Look, no technology today or later nor the propaganda on an internet messaging service, or even a billionaires underinformed or warped sense of climate can change that outcome. It's a physical law made by an authority more powerful than all of us combined, nature. I'll explain more later. For now. Please don't panic. We really can fix it now that we know exactly what's wrong! But it does take time to mature trees and reestablish our Titans.

47.75 PPM MAKES CH_4 EQUIVALENT TO CO_2 PPM.

I mentioned this, but it deserves a better definition.

In addition to having and making too many blankets we have other issues like methane (CH_4). Methane is like CO_2 except it's much worse in its natural form. It's related to and called "Natural Gas." It's used everywhere to heat and cook and as a fuel goes it is a good wood substitute. After its burned it becomes CO, NO_2 (nitrogen oxide), and CH_4. It's not good to have it in natural gas form within our atmosphere either. It doesn't belong there before or after burning. The CH_4 component is more effective as a greenhouse gas in creating the global warming blanket. In fact, it's 25 times more potent than CO_2 for retaining heat in the atmosphere. As a contributor in the fifth parts truth, more diligence is required. What CMS understands is that forestry harvest sites are contributors of CH_4. Therefore, CMS's residually decreases CH_4 by ending clear cuts and other bad ideas in forestry. But CMS is limited in impacting CH_4, but it does none the less. In fact, CH_4 is the prime reason for pursuing human emissions reduction.

In July of 2024 CH_4 in Earth's atmosphere is 1911 ppb (parts per billion). I equate ppb into a CO_2 ppm equivalent. I do because CH_4 goes unadvertised, and it really should be included in the publics global warming knowledge. CH_4's CO_2

equivalence is another 47.75 ppm of CO_2 in our atmosphere. So, technically our current 427ppm CO_2 blanket is really at 474.75 ppm CO_2 equivalence (as of July 2024). Yep, more than a double blanket but worse because CH_4 is also flammable.

CH_4's climb since 1750 CE has been a bottle rocket. Since then, CH_4's atmospheric content doubled and continues to increase in volume yearly. Yes, some of it can be correlated with CMS forestry data, but not all of it. Which is also another warning. CMS can obviously do something about a lot of CH_4. The warning is it can also be correlated with rising ocean temperatures and fossil fuel use. Oceans might be the number one source of CH_4. Other sources include uncapped fossil fuel wells, leaky pipelines, natural gas use, and refining oil. And that is why I said its prime for a human made emission reductions, because fossil fuel use might make the majority of it being released. But maybe not.

Oceans hold frozen layers of CH_4 as clathrates in their sedimentary bottoms. Basically, stockpiles' of that flammable, atmosphere killing stuff is all over the deep oceans' layered bottom. Usually at 3,600 feet, 1,100 meters or more below sea level where cold temperatures and pressure keep it at bay. Unfortunately, those piles are only a couple of degrees Fahrenheit from being released into our atmosphere. At less depths, with warmer conditions their thawing and releasing CH_4 now and have been for eons. The same goes for the frozen tundra in the artic. Thawing tundra is also a source of CH_4.

None of this is new data, but ocean and tundra warming has made CH_4's release an ever-increasing volume since 1750 CE. And so, what, right?

First, it proves CMS's precedence. 1750 was long before fossil fuels and yet sequestration decline had already established warmer temperatures. That warming allowed recorded CH_4 releases to increase. How about that, an unexpected proof of sequestrations demise.

Second, and the downside of being right. If ocean and tundra stockpiles of CH_4 clathrates release, we as a species are done for, period. That is the end of us and all other fish and animals on Earth. Only a couple of higher degree's difference in temperature before nobody is left to pay the bill. Our atmosphere would become either flammable or CO_2 saturated from the burning CH_4. But changing forestry with maturity also fixes this dilemma as a residual. Literally, forest maturity fixes it by shedding blankets. That allows our oceans and tundra to cool, so we and polar bears can again sleep better. One more thing, a couple of degrees of ocean temperatures is the difference between here and not here, really?

Time for a disclaimer. The ocean depths keep their temperature as a physical constant, 28.4° Fahrenheit, -2° Celsius. Therefore, I'm not completely sold on the idea that the entire ocean temp at 3,600 feet, 1,100 meters or more in depth can change anytime soon. I really mean in millennia. However, global warming affects oceans that are not that deep and tundra not that far into the artic. Well, heat generates more heat. So, I mentioned it. This is one of the CH_4 topics CMS implies as hypothesis. I do not say it is a fact if it involves CH_4 not tied to forestry. CH_4 and the ocean and tundra are not the focus of CMS studies. What I do know is clear cutting releases a lot of CH_4 that drastically adds to the *carbon hump*. And forest maturity does cool Earth. Hence, CMS has an impact but I'm not ready to claim any more than I have already theorized. I'm not the expert oceanic or tundra CH_4 releases need.

That said, CH_4's levels increasing since 1750 as a proof of CMS results is all I really need to convey.

MORE PROOF OF CLIMATE CHANGING CONDITIONS

The world's population of 20 million people around 5,000 BCE is around global warmings official start. That population level doubled 2,000 years later. Over the next 3,000 years, we successfully increased ourselves almost six times to 232 million people. That was probably humans most significant advancement as some believe we became too big to fail. Or did we actually become big enough to fail atmospherically? That growth happened before the year 1 CE. Now, 2,024 years later and Earth has around 8.1 billion people and *forestry demand* has spread across the globe. What can be said is it took a lot of humans a long time to break sequestration into the shards of old-growth and mature Titans that still remain. Population increasing definitely provided the labor needed. As proof, atmospheric ppm's levels increasing and CO_2 outflows diminishing over the same time do kind of show a population correlation to global warming. That correlation is difficult to argue against, even for me. However, that correlation shows itself as a measurable symptom of global warming, it's a variable not the specification. Population provided the labor and demand, not the direct cause. This book has graphs that show this better.

POPULATION INCREASES DO INFLUENCE GLOBAL WARMING WITH FORESTRY DEMAND BUT IS NOT THE DIRECT CAUSE. IT'S WHAT SOME OF THOSE PEOPLE DID BADLY THAT CAUSED IT, FORESTRY MANAGEMENT.

The geo-engineering used was the improper and inefficient use of forestry resources during past and present human self-domestication. This well-pronounced effect is measurable within current tree or forest ages when considered as not having been harvested to their current impeded and absent CO_2 sequestration ability.

The difference defined by human forestry absence dictates the level of an *impeded fast-cycle CO_2 sink*. As in how much it is impeded. So, when you compare a replacement tree's sequestration ability now to its potential, as measured as not being previously harvested or by human absence, the result can be millions of percentiles of impedance. That is because trees can live thousands of years and the fact that maturity is *proportional* to sequestration ability. In addition, atmospheric inflows and outflow ppm delta's provide proof over time, just as the following statement provides added math and CMS logic precedence.

For example, currently the tree absorbs 40 Lbs. of CO_2 annually, but if that forest had been managed by natural attrition or never clear cut, today a tree within it could be absorbing 3,000 Lbs. of CO_2 annually. It is impeded by (3,000 Lbs. CO_2 x years alive (3,000)) divided by (40 Lbs. x number of years alive (20)). Or roughly for this example, about 2 million percent impeded. The immature tree is an *impeded fast cycle CO_2 sink*. Yikes! Now, let's drop the logic bomb on the uninitiated.

Therefore, if you remove humans from their historical forestry use, but keep human and natural CO_2 emission levels; there would be no such thing as CO_2-driven "climate change" or global warming occurring today. Global warming conditions become impossible to form as a result of having adequate forestry sequestration globally.

Even true on Earth's rather limited forestry available. It is all made a fact by tree maturity being proportional to its sequestration ability. And CMS's *proportionality* in general. For example:

If some human's had not impeded our global forests' maturity levels, their sequestration capacities could absorb well over the 435 plus gigatonnes of CO_2 emitted annually. That is the natural order of sequestration and unlike today's geoengineered scenario.

I'm using rounded numbers and not showing my math to make this point.

[1] 70 mature trees per acre at 571 Lbs. CO_2 per year is around 40,000 pounds total CO_2 sequestered from the atmosphere annually per acre. We're not using old-growth, just mature marketable trees around 62 years old from a not-so-great region and species, to be conservative.

   a. 10.3 billion forest acres remain on Earth. Times that by the 40,000 Lbs. per acre. That's 187 gigatonnes CO_2 capability, annually if matured 62 years. Because of the *carbon hump* that is in addition to today's sequestration ability. Now add this in, in another seven to ten years that number can double to 374 gigatonnes per year. And again, in 7-10 more years it can grow to 748 gigatonnes. Wow, right? That is forestry's sequestrations power. It continues to grow each and every year it matures.

[2] As an old-growth forest, which can be anywhere from 100-800 years old and thousands of years older is even possible. Old-growth can have 7-10 times more sequestration power than the 62-year-old example tree. It of course depends on how old, location, and species. Being conservative again, old-growth can sequester between 1,309-1,870 gigatonnes CO_2 annually, and it keeps growing larger every year.

   a. Keep in mind I'm using a perfect world but conservative projection that uses the entirety of global forests. It's conservative enough to make it difficult to credibly argue against. But sloppy enough to ask if this CMS point is truly undeniable. My answer is, it is accurate in assessment but lacks a well perfected precision. Therefore, yes, it is undeniable proof of forest maturity's sequestrations power. Although precision lacks in the estimate it still hit the bullseye. Where it hit in the bullseye is moot (precision). Plus, results are reproducible. Oh, and this point is made in other ways as well.

START WITH THE EMPIRICAL AND THEN APPLY LOGIC.

It's by far more natural for Earth's forestry CO_2 sequestration capacities to greatly exceed all CO_2 emission sources, exponentially. When and from where CO_2 is delivered to photosynthesis doesn't matter to the forest's natural order of operations. Plants in general don't care about the CO_2 supply chain or when it shows up. They do care about the lower ppm limit in photosynthesis which is around 200 ppm. It can't get enough CO_2 below 200 ppm to grow. The upper limit is negligible to the amount of available light and temperature limits, and that is happening already. Which is a problem. I address it as excessive CO_2 fertilization in the next section and photosynthesis limits later.

For now, add that atmospheric residence conditions apply time before sequestration happens. That added time serves to level the outflows of CO_2 from the atmosphere into photosynthesis. Which is due to photosynthesis's randomized supply and randomized use of CO_2. This randomization accounts for more atmospheric residence time. Residence time therefore acts like a faucet in a dam. A dam constructed by weather and chaos. For example, the release of CO_2 from a factory lingers; that molecule may not see photosynthesis's sequestration for decades or like today, never. It stays within atmospheric residence instead of coming in contact to a plants intake. Even if the weather pushes it into a forest the chance of being absorbed increases but it's still not guaranteed, it's a random act in chaos. To be sure, it can eventually happen under normal conditions, but Earth is far from normal today. So, the CO_2 reservoir behind the dam grows and stays to form the fifth part, our CO_2 blanket. While the faucet shrinks.

Ever wonder why its hotter in the city than the country? Resident conditions contribute to that effect by holding the CO_2 blanket the city creates over that city until the weather moves it elsewhere. The city creates a microclimate by emitting more CO_2 than it's sequestration can absorb. It is in the *sequestration dependent carbon hump* it made. Which comes down to this. Even though the Americas' forests' could take in more gigatonnes then emissions produce they don't get the opportunity to do so. There is no pipeline moving the city's emissions to the forest. Especially with today's immature and impeded sinks. These days, CO_2 is waiting in a queue. Which brings me to another reason that is even more powerful.

To the point, photosynthesis evolved to take whatever CO_2 it can get during the daytime. Yes, it's solar powered. Here's the empirical fact to remember. Plant's evolved to process way more CO_2 than is normal. It's to their determent and not in their benefit. The randomization and chaos in supply required them to do that. They did that so they could get enough CO_2 to grow even when CO_2 levels approached the lower limit of 200 ppm. It's a safety feature; remember they must grow, or they die. Unfortunately, their evolution didn't evolve a regulator for too much CO_2. Which is what is happening today, too much.

Excessive CO_2 fertilization is occurring globally. Everywhere! The randomization and chaos no longer reign supreme, the dam has over-topped because humans closed the faucet. At .044% of the atmosphere, 427 ppm, there is way too much CO_2 being presented to plants, so randomization is circumnavigated more often. Today, too often.

Note: I'm not saying trees. Be assured they are also affected but this problem is much larger than just forestry. CO_2 is exceeding the normal needs of plant life to grow. To be sure, plants absorb more CO_2 than needed when they gain more exposure. Yes, that also means they could absorb even more than they do now, but not without even more permanent damage. CMS explains why Earth is currently opposite of the natural order of CO_2 sequestration operations by exposing excessive CO_2 fertilization for what it is. An *accumulating decline* powered by the abyss's maturity vacuum.

EXCESSIVE CO_2 FERTILIZATION.

Another unfortunate precedence of CMS. I have seen interesting, creative, and opinion-based work saying excessive CO_2 fertilization is not a terrible thing. I disagree. Negative effects outnumber the positives in growth stimulation and less water used to grow.

First, plants and trees grow too fast and too high to support themselves and fall over. I have a spruce in my backyard and other tree species in the neighborhood are proving that point. **Second**, food loses flavor because it grows too quickly and too weakly. It's not the older person's taste bud declining to their memory of better tasting fruit from an old tree they've eaten from, since a kid, like me. Taste buds don't decline unless your brain does. These days, we really are consuming empty calories. **Third**, plants don't retain or use the amount of water they should. They become flammable sooner in the summer. Gee? Why do we have so many wildfires these days. Yea, there is more than one reason, but this is one for sure. **Fourth**, one of the stranger effects; the green plant or tree you see today is a brighter green than one a hundred years ago. Weird right? The world is a lighter shade of green because of this climate changing condition's impact. **Finally**, you can see this next negative effect every spring for yourself. Although this is an accelerated result. Have a look along any busy roadways. You'll see yellow, sickly looking but quick growing grass and weed stalks in the spring. Their lighter in color and weaker in structure because vehicle exhausts blast them with excessive

CO_2 and CO fertilization. Even though that yellowed sick looking plant grew a foot or more quickly it dies within days of sprouting. That is only some of the symptoms of excessive CO_2 fertilization and why CMS calls it bad. It is an *accumulating decline*.

To explain further, photosynthesis is mostly undefined by species in an upper ppm level associated with excessive CO_2 fertilization; other than to recognize the across-the-board damage. We do know it occurs and is a species influenced effect. As in some plants succumb to CO_2 poisoning faster than others. Just as humans die from oxygen toxicity plants can die from CO_2 toxicity. At that scale, it will be too late to do anything about global warming, checkmate. However, this can be fixed with maturity spread across billions of acres knowing residence time's randomization will dilute excessive CO_2 in healthy, mature microclimates. If you have ever seen the baren Earth in Road Warrior movies, that is the what the result of excessive CO_2 fertilization looks like. Everything is dead or dying.

For now, I have to admit excessive CO_2 fertilization can help CMS mitigation efforts accelerate sequestration's recovery. But at what cost? That I do not know yet because the question is moot. We have to start somewhere even if that start is well pronounced and documented excessive CO_2 fertilization occurring globally.

I HAVE SOME EXPLAINING TO DO

WHEN DID GLOBAL WARMING BEGIN.

Human self-domestication began millennia ago, and with it, so did global warming. It all started when humans settled into fixed places, adopting disciplines like animal husbandry and farming instead of remaining nomadic. This was also when we began harvesting trees. Over time, we became more adept at utilizing forestry resources. We had used wood as fuel for eons, and still do today. Eventually human's learned to build everything we needed from it. Essentially, humans figured out how to exploit forestry. This culminated around 8,000-10,000 years ago, coinciding with the noticeable growth of human populations in permanent villages and farms dotting the landscape. That permanence in our location is when *demand driven forestry* began by applying our needs to *convenient forestry*. That need hasn't stopped.

Back then or now, humans didn't use forestry as effectively as we thought. We believed it was self-renewing and didn't require human care. We've learned much since then, but even CMS doesn't have all the answers about managing such a diverse resource. Nonetheless, we've continued to use it for millennia without much in the way of foresight. The key difference now is our understanding of maturity, which helps significantly by adding more knowledge to stewardship.

USING FORESTRY ISN'T THE PROBLEM.

The problem is how humans cared for it. Today's stewardship efforts are dated because we now understand that trees hold the *binary restricted resource* sequestration. Believing the production of woody biomass was the only resource in a tree is where our stewardship fails miserably. As

CMS can now point out, trees aren't the renewable resource human's thought they were. There are 2 resources in the tree, and one hides the other. *Sequestration value* is hidden by Humanities partiality towards biomass reproduction and its immediate usefulness. Call it our continuous need for instant gratification.

We've touched on Humans having "successfully" modified our forests to suit demand. There's that word again, "demand." Humans want or need it, so humans get it. By CMS definition that becomes *demand driven forestry*. Making forestry answer our demands via human engineering is considered an achievement in human self-domestication. And it is… sort of. Unfortunately, that achievement has strings attached to global warming.

One very strong string attached provided humans with the ability to *impede fast-cycle CO_2 sinks,* sequestration. I said sort of earlier because the CO_2-*sequestration value* of a tree is not an easily renewed resource. The binary part, biomass, renews exponentially faster than its restricted counterpart, *sequestration value,* can. That fact is magnified by the time required for the tree to mature and become net carbon neutral or negative within a human made *carbon hump* tied to forestry's use. You see, all wood products leak CO_2 initially and then over time because nothing is carbon free when human's use it, we are *emissions dependent*.

FIGURE 4: A GRAPH OF CO_2 PPM, SEQUESTRATION FAILING AND HUMAN EMISSIONS INCREASING.

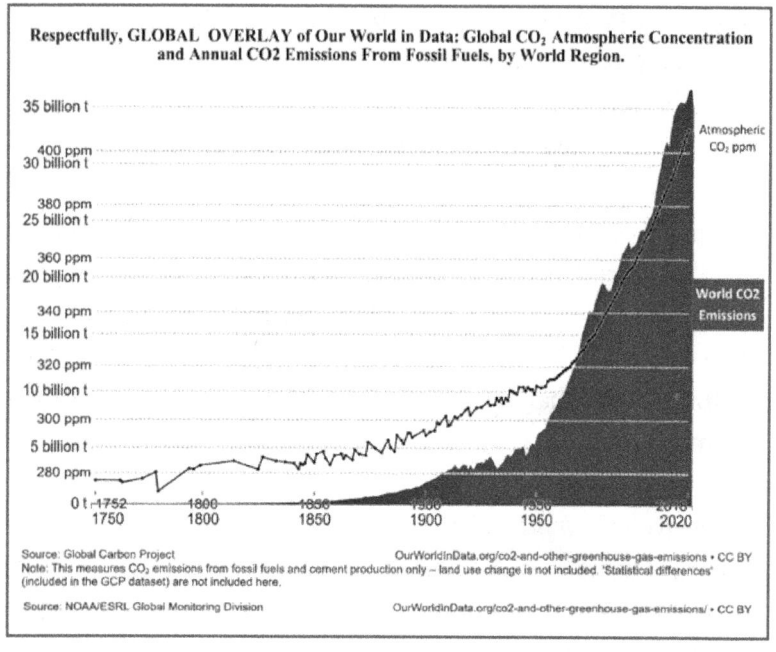

Figure 4, an overlay of atmospheric ppm and the worlds emission levels. Note: 1850 CO_2 ppm, 1950 ppm, and Human CO_2 Emissions. Overlay is 1750 and 1752-2020.

Figure 4 clearly shows Earth's CO_2 ppm levels increasing over time. The ppm increases initially took eons to establish before accelerating into today's rocket ship like trend. No emission sources correlate to the thousands of years of documented ppm increases nor do today's obscene human emission levels. Not solely, and not without sequestrations presence to balance the atmospheric equation (referring to CMS's Sequestration Computation). Nor can the ppm increases be explained, nor can their occurrences be reproduced without CMS. That sign of something other than emissions having caused the historic and modern ppm increases is CMS's undeniable conclusion. Figure 4's overlay of data sets shows, clearly, exactly what caused global warming, sequestrations decline and nothing else. It begins and ends at the same place, human engineering.

This book's small format makes some things difficult to see in Figure 4, but you should be able to see the rise of ppm that showed climate change existed long before human CO_2 emissions took off. You can also make out the 1850ish datum and the 1950ish one as well, which we explain later.

The figure also shows the demise of remaining old-growth 1950-on. There can be only one conclusion. Sequestration was being affected and the emissions building Earth's *carbon hump* was part of the problem, but not all of it. In fact, not much of it at all.

Okay, so the ppm and emission lines may look similar but they're unrelated. Don't believe me? No problem. I initially saw the same similarity between the two, so I went deeper. Have a look at Supplemental Figure S1 in the back of this book, page 229. It provides a clearer and longer-term ppm scale with added explanation.

Have you looked at Figure S1? Good. Rising ppm and no emissions, hmm? What do you think now? Well, how about I say rising ppm and declining forest maturity? Yes, that is exactly what it shows? Both Figure 4 and Figure S1's ppm levels correlate to forestry demand. The two figures act like they were traced from historic forestry use.

We don't need a big correlation graph to make this point. It's too big for this book's format. But that's okay because Figure 4 and S1 speak volumes all by themselves.

Another point to consider, around 1850 there's a unique sequestration signal in the ppm data. It's the first year the global *CO_2 fast-cycle* gets out of balance with emissions and begins signaling a serious problem, global warming. That negative indicator worsens from 1850 on. You can see the *fast cycle sinks* in forestry shrink in numbers and ability. It's the up ticks and down ticks in the ppm line as they decrease in amplitude (their spread) from then on, until they literally disappear around 1950 (at this scale). None of that disappearing fast-cycle or growth cycle pattern relates directly to emissions.

Do you see that moving-sideways ppm line before 1850, and the upward lunge before 1800? That's volcanic activity, and I mean a lot of volcanic activity, occurred then. But that also means sequestration levels could keep up with it! More coming on that later. For now, it creates a lot of chatter that needs filtered out to spot-on the 1850ish *climate change datum* time. Chatter that makes me say 1850ish and not 1850.

Next, **Figure 5**. Time to zoom in to an annual fast cycle CO_2 sink cycle or more commonly known as Earth's plant growth cycle.

FIGURE 5, MAUNA LOA CO_2 PPM MEASUREMENTS.

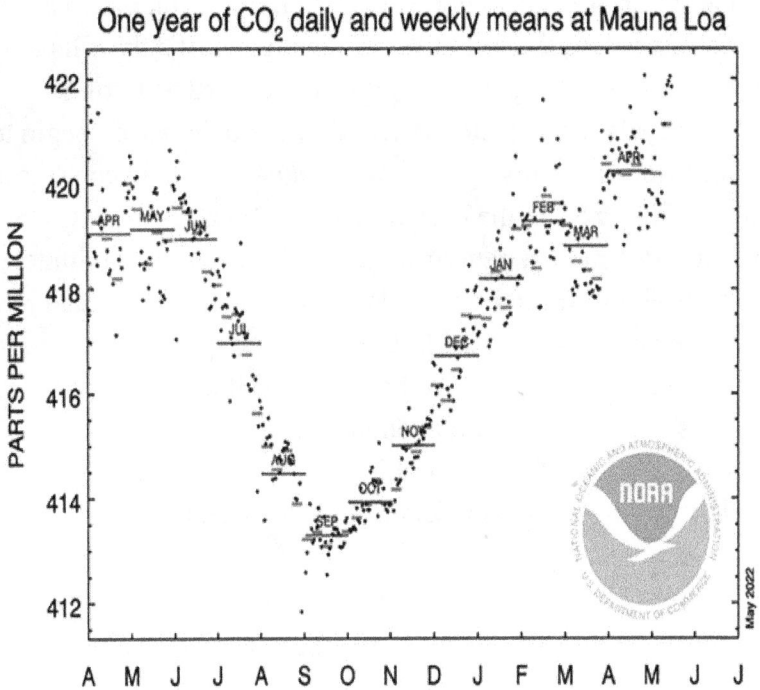

Okay, let's discuss **Figure 5's** up ticks and down ticks. Those are a typical fast-cycle sequestration pattern measured in CO_2 ppm. This is a snapshot of our planet's growth cycle. This should make it easier to see the sequestration signal decrease and then disappear from the 1850s to the 1950s in Figure 4. Figure 5 is Earth's growth cycle in 2020-2021, published in May of 2022; The source is Mauna Loa observatory's CO_2 ppm measurements 1958-current. Figure 5 shows 16 months of the typical fast-cycle sink/growth cycles of Earth.

The black dots are the CO_2 ppm readings taken during the weeks in the month shown by its first letter and during 2020-2021 growth cycles. Note, the ppm ends a lot higher than where it started. I'll be talking about that soon enough in the 1950ish datum section, unfortunately.

An average CO_2 atmospheric growth cycle within the fast cycle sink pattern is an uptick in CO_2 during winter, then a downtick before the end of the following fall. It's the hills and valley you see in Figure 5. That pattern existed as Earth's growth cycle long before humans did. It did, however, begin to be influenced negatively around 6,000 BCE, 8,024 years ago, also in Figure S1 in the back of the book, page 229. By CMS's 1850ish datum, that pattern started to shrink (lose amplitude) and has almost disappeared today. Unfortunately, the trend has gone negative twice in the last seven years; first in 2016 and again in 2019. The evidence for this is available as "CO_2 ppm Delta's" in the CMS journal on the maturetrees.org web site. It's also too large for this book's layout. From 2022 to today, well, it is not getting any better. And won't on its own.

ULTIMATELY, WHAT CMS POINTS OUT IN PPM DATA IS NOT A GOOD THING FOR US, NOT GOOD AT ALL.

Back to the naturally occurring CO_2 sink cycle's graph in Figure 5. After winter, spring arrives, and the natural sink cycle should decline ppm levels during plant growth periods, meaning spring through fall, which happens every year in a big ppm swing, or at least it used to and should. And could again with CMS's help. In a perfect world, and if sequestration outflow was healthy, the ppm line would be more of straight horizontal line with CO_2 inflow equaling outflow. Resident time would create a valley but nothing like what's viewable in Figure 5. Also, the valley gets higher on its right side every year. Indicating, not enough sequestration occurred so more CO_2 becomes locked in our atmosphere. But that's only part of the problem, because the valley does indicate sequestration still works, just not like it could or should. Year after year the sequestration amount declines and has since humans started harvesting trees.

Figure 5 gives a perfect example of the growth cycles' upticks and downticks. But Figure 5's 2020-21 swing is not viewable in Figure 4, because the graph's inflows and outflows disappear in the ongoing buildup of CO_2 ppm in atmospheric residence and that other reason: Shrinkage.

You can see sequestration shrinking in amplitude 1850-1950 and continuing towards disappearing after that. In later years, we must zoom in to Figure 4 to even see the cycle because of the year- to- year attenuation gathering in sequestrations signal. That is shrinking over time. The sequestration signal is going away because it is becoming weaker. And that's a very serious problem. A huge climate changing problem that emission science missed. A mistake I explained earlier, as believing CO_2 atmospheric outflow was static on page 46. Well, science is wrong about that to say the least and now we might have to pay for that mistake with all Humanity.

That said again, do you see the tiny downtick from February (F) to June (J) on the right-hand side of Figure 5? Good, here's what that's all about. Earth has two hemispheres, so we also have two growth cycles created by our tilted axis wobbling, which is called precession. That wobble happens as we travel all the way around the sun, one year. The small hill and valley is the southern hemispheres growth cycle. **First**, the cycle is unhealthy and shrinking, so it's a small hill and valley. **Second**, it's a small growth cycle. Only 32% of Earth's entire landmass is in the Southern Hemisphere. So, what you're looking at is largely what's left of the Amazon's old-growth's sequestration doing its best to keep up, but it obviously can't. I would have mentioned the other Southern Hemisphere countries, but there isn't enough old-growth or maturity to really measure. But I must!

Specifically, the Southern Hemisphere downtick is created by roughly 90% of South America, which is mostly the Amazon. And the southern part of Africa, and the Australia/New Zealand sink cycle. Yes, there are a few large and small islands there as well. However, their combined growth cycles don't account for much sequestration, not anymore. They're all thoroughly logged out and lack maturity. Again, check Google satellite maps and tour below the equator to see for yourself.

Before 1850, the cycle's trend at the end of every fast-CO_2-sink cycle (by fall) would be close to equal to even below where the CO_2 ppm level began at the end of the previous winter. Meaning a lot more sequestration was happening before 1850ish.

To give you a general summary of the idea, the ppm outflow trend is around $y = 9.56$ CO_2 ppm difference in the amount sequestered from the atmosphere from 1974-2021, 47 years. Each ppm equates to around 7.82 gigatonnes CO_2. So, 9.56 ppm is 75 gigatonnes of CO_2 (averaged) more than Earth's sequestration could deal with and so it all ended up in atmospheric residence during that timeline. Today, a scary percentage of less sequestration is available globally when compared to before the 1700's. CMS estimates it is around 30-36% less. I don't have an exact figure due to data processing limitations; however, there is no doubt it is within that range.

CMS was also able to predict Earth's ppm level from May 2022 to May 2023. That increase was 2.8 ppm and equivalent to 21.9 gigatonnes added to Earth's atmospheric blanket. An almost 30% increase in one year in comparison to the previous 47-year trend of $y=9.56$ ppm, 1974-2021. Failing sequestration yes. Emissions increasing? Also, a yes. However, failing sequestration is the bigger impact of the two. We know because of the CO_2 deltas (the difference between atmospheric CO_2 inflows to out flows, the sequestration level is outflow). The delta amount, the CO_2 amount sequestered, decreased every year. But don't despair; there is another way to read this part and that is with hope. Finally! Some hope!

HOPE IS REAL AND PLENTIFUL

The hope comes from sequestration's balance point with emissions (sequestration computation is all about the balance point). Looking at 2023's information and the emissions and sequestration balance was obtainable at around 21.9 gigatonnes. As in 21.9gt needed sequestered for the balance to occur. In considering global warming's obtuse gigatonnes in measurement these days, that amount is not that much. However, I have to rain on the picnic a lot. We still have that very inconvenient 1,541gigatonnes in atmospheric residence to deal with. Therefore, complete mitigation is around 50ish gigatonnes per year and requires 54 years to end global warming. Which is hope. But we can do much better. At 100 gigatonnes sequestered annually mitigation is just under twenty years. Seems doable, doesn't it? I mean any mitigation plan CMS suggests can do that with scale and do so in 20 years. Which I believe provides a lot of hope. However, the ongoing annual ppm increase makes it pretty obvious we don't have time to wait and see what happens.

Just one year later, May 2023- May 2024, an average 2.9 ppm increase was recorded. So, ppm again increased over sequestration balance by adding another 22.68 gigatonnes to Earth's atmosphere. Breaking that down, 2023's 21.9gt is now 2024's 22.7gt to reach the balance needed. Yikes, that's a 4% increase in one year! Normally, that would not have been done easily, considering we are measuring the entire global atmosphere. But also consider emissions are rising and sequestration is decreasing. So, that 4% increase was unfortunately very easy for humans to accomplish. So, we get the global warming trophy again, this time for speed exhibited in our own demise.

Look, prior to 1850's there was almost enough sequestration ability to deal with natural and human emissions, combined. Even with today's ppm increases! To break that down, in 1700 CE there was just enough sequestration to hold back global warming. Prior to 1600, there was more than enough. In fact, before 1600, sequestration worked "well enough" that climate change would have taken almost a thousand years to show up as it did around the 1850ish datum.

CO_2 ppm trending up, up, and up again each year while sequestration is decreasing didn't make it hard to pick a starting point of a **global climate collapse** beginning around the 1850ish datum. Yes, **global climate collapse**. That's what really began around the 1850ish datum. Sequestration shrinkage is creating all those *accumulating declines* and that confirms the collapse has begun. How long will it take? 840 ppm rings that bell. The concerning problem of today is it all continues to accelerate faster and faster every annual growth cycle that forestry CO_2 sinks are impeded. And yet emissions reductions get in the way of fixing that while creating even more problems for sequestration to fix.

A HUMP MADE OF CARBON DIOXIDE.

As to the *carbon hump*, it takes up to 30 years and as little as 10 years, occasionally, after a clear-cut and replant of acreage to sequester more than what's being released as emissions by the land, waste created, energy used, and wood products made. Those are all CO_2 and other types of emission sources that are mostly unaccounted for in the human forest relationship.

Sequestration returning to value can take up to 30 years because the hump is a slowly exploding bomb. It can take years for the harvested forest land to mature. Mostly, its burned or left to rotting. Both release CO_2 and other emissions in quantities. It takes even longer to become net CO_2 neutral if natural regeneration (allowing adjacent trees to seed the clearcut part) is used to regrow the tree stand. Natural regeneration doesn't cost money, so it's the preferred global afforestation method. Although it does require more time to regenerate the biomass, the expense in doing so does not require any money up front like replanting does. Replanting is expensive for reasons like seedling natural attrition, the cost of saplings, and labor. So, it's not usually done willingly by unsubsidized timber producers.

The next part you know from earlier in the book but it's worth mentioning again. Now you can add in the *carbon hump* that explains why I used a thirty-year-old tree for the demonstrations. Trees younger have zero atmospheric impact. Which makes maturities impact even greater. I know, but its true. And one of the causes of global warming CMS addresses.

I love repeating this and adding to it each time:

During a typical thirty-year-old tree's annual growth cycle (spring-fall), it will absorb, use, or sequester from the atmosphere up to 163 Lbs. of CO_2. Now, the crucial part, at 72 years old, it can do the same to 1,100 Lbs. of CO_2. Quite a difference, right? With just under three trillion trees on Earth lacking maturity, maturity in forests really matters!

Okay, maturity increasing a tree's sequestration ability is both incredible and empirical in measurement. So, it's also undeniable. It is the most powerful CO_2 controlling weapon ever to be inspected by Humanity. Now we just need to buy it. It's not time to celebrate either, not yet. We do need to work on a problem to use that information and cure global warming for good. And dare I suggest this, "control Earth's thermostat and manipulate our weather, eliminate drought, improve our self-domestication with advancement, oh, and decrease the costs of weather driven natural disasters, reduce fires, plus make more productive and better forests!" I'm all for it. No downside is detectable when using a natural fixture to fix our problems. Time to bring on the Titans is now.

BRING ON THE TITAN'S

Here's the issue with that tree maturity proof I love repeating: sadly, a 30-year-old tree will likely never reach age 45 under today's *demand-driven forestry* practices. This reality has nearly bankrupted Earth's best CO_2 sink, forestry. Trees are typically clear-cut between the ages of 20 and 45 to meet the demand for forest products, often even earlier on privately owned lands. Unfortunately, most forests are privately owned, making it rare for any tree in a marketable forest to mature beyond 45 years. In certain regions and hybrid species, trees might even be harvested as early as 17-20 years old.

Economic downturns, like recessions and depressions, can temporarily increase the average tree age. Within some government-managed forest land in the USA, trees can reach over 45 years old regularly. However, generally, trees mature only to 20-45 years, despite having the potential to live for several hundred or even thousands of years. Granted, some trees are allowed to mature to 45-60 years or even older, but these are exceptions with a catch and not the rule.

Most trees that do reach 45 years or slightly beyond are typically found in less productive growing regions, where their size isn't marketable until they are 45-60 years old. But surely not all forestry falls into the timber reproduction category, right? Some trees within national parks are protected, but this accounts for only 5-7% of the world's forests at best; more realistically, closer to 3%, though data inconsistencies prevent a confident figure I can read between the lines.

I wish I could say that global forestry doesn't fall into the marketable category, but that would be a lie; one often repeated by less credible sources. Speaking of credible sources...

What is an Old-Growth Tree?

My definition of an old-growth tree and the United States government's convoluted definition(s) are not the same, not even close. Although I do appreciate the government's old-growth definition could protect more trees and increase maturity, that is not their intent.

If a tree isn't ancient within its species average lifespan it isn't and can't be considered old-growth. Period. But calling immature trees old-growth sure makes public distributed government charts appeal to our environmentally concerned population, so our diligence is now required, it's a trick. Unfortunately, government practices seem intent on hiding the lack of maturity in our forests. The government's mindset is being driven by *constrained deforestation*'s outcome and hiding that seems to be their first instinct. All so they can continue *demand driven forestry* practices. Practices political pressure is applying on good people just trying to do their jobs.

The fact is, the United States doesn't have much old-growth remaining, very little. Today, useful but highly insignificant science is being used to conceal that fact. They are throwing so much insignificance around the actual lack of old-growth fact and advancing their unspoken agenda it's exceeding the normal bureaucratic nonsense.

So, my government, what's wrong with just being honest about the problems we and you must face, immaturity in our forests, biomass efficacy, and *sequestration value*? Hiding it isn't going to make it go away. Ah, but you hope to get your way, the old tried and true *demand driven forestry* way, don't you? You should not do that, not after understanding sequestration science and with the truthful old-growth definition you can't. So, you decide to make your own old-growth definition to hide your predecessors incompetence. How disappointing.

I SUGGEST, AN OLD-GROWTH TREE IS ONLY OLD-GROWTH WHEN MATURED TO BEYOND ½ OF ITS SPECIES AVERAGE LIFE EXPECTANCY. A LIFE EXPECTANCY WITHOUT BECOMING HARVESTED. TREES ARE ONLY MATURE GROWTH AFTER ONE-THIRD OF THEIR AVERAGE LIFE EXPECTANCY.

The UN and most scientist's define an old growth forest as, "as naturally regenerated forests of native tree species where there are no clearly visible indications of human activity, and the ecological processes are not significantly disturbed." PERIOD.

The government's attempts to change that definition include the use of tree canopies, forest density, tree size, and many other insignificant reasons to decide what old-growth is. Those side shows to defining old-growth are avoiding the use of current age to life expectancy and human interaction. And the real descriptions of old-growth were made fact long ago with humans becoming intelligent. And by that fact and the one I've suggested all old and mature growth should be governed, or else. Maturity first, all else is moot when it comes to any tree's age.

By fact holding definitions and not the current attempt at a fairytale, the United States has less than 3% old-growth trees remaining. And nothing like the current FS-1215a, a report telling the public we have over eighty million acres of old-growth. That is just not true, not anywhere near it. Its immaturity propped-up to make it into something it's most definitely not. That report never mentions life expectancy to the tree's current age to determine whether it's an old-growth or not. That makes our diligence even more necessary. And sequestration science's disappointment rabid. Because of their smoke and mirrors, their intent.

Why? Things like logging impact, global warming, *accumulating declines*, and drought all manipulate the government's newest old-growth criteria. That "make believe" criteria can only result in miss classifying trees that are just getting over the *carbon hump* and somewhat mature stands.

I can hear them now, "this stand does not fit the old-growth criteria we made up so it can be harvested." So, why would they want to do that? By changing the definition, they have legally unlocked them from laws protecting them, all so they can be logged, again. As you'll discover soon, most government managed land contains that misclassification partiality towards harvesting trees. That "new" old-growth definition will make them harvestable by working around hooty, the endangered species act. In addition, the FS-1215a's incorrect classifications will falsely tell the public stories about the millions of acres of old-growth we have, that the government manages, which implies logging them is no big deal because we have so many. Theirs is an effort in the unrequired and truly unnecessary shoveling of malarkey upon the public's belief. Why is it malarkey? **First**, because it's deceptive. Disguised as old-growth protecting conservation, it is anything but. A deception spread across entire organizations. Its "group think" if it does not account for sequestration value or tree age. Both things require maturity. And may not produce under their new criteria's listed attributes for centuries because of *accumulating declines*. And they know that. **Second**, because they don't know sequestration science or natural attrition harvesting. Those facets of sequestration science end all the problems government forestry faces without the deception. It's a no brainer, which does require having a brain that doesn't join in with peer pressured deceptive practices.

Yes, that is an opening salvo in a battle I can't win with facts. Its point is too rogue to their intended purposes to ever be heard. I again suffer my grandfather's curse. Unheeded yet again. So now it's playing out in global warming's advantage. No profit foreseen to feed my buzzard as we all await fate's dark cloak and boney fingers to arrive. Those arrogant Idiots will kill us all as they watch humanity struggle from their grouped together cheap seats. A stadium filled with stupid, all so there's a chance for their career to advance. It won't, because you caved when you should have delicately fought.

It's also my way of foreshadowing the problems with data available on sequestration. Data is not available, or it's manipulated so badly towards demand driven practices it must be sorted carefully.

TIME TO HIT THE DATA PILE AND TAKE OUT THE TRASH.

In fairness, the USA's Department of Agriculture, USFS and Department of Interior, BLM and the United Nations's global data is helpful in untangling fact from many fictions about forestry. But not as directly as one would think. They are not the only CMS sources; they couldn't be for many reasons, deception being the main problem. Basically, they are forced to make-up, use and pass manipulated terminology and data. But not always. One must always keep in mind where statistical data comes from, or from whom, meaning the BLM or USDA's Forest Service or the United Nations didn't count the trees or do all the surveys themselves, they sometimes relied on landowners to supply data, big corporations mainly.

Okay, the verifiable truth is this, 85-95% of forest acres globally consist of marketable trees harvested before age 45. And very little old-growth trees exist globally. Maybe 3% of 3 trillion. But marketable status is probably higher and around 97% because most forest land globally is privately owned and controlled for biomass production and not for the public and not for *sequestration value*. Yet.

For example, to build my own confidence with those statistics I use 85% of global forest land to calculate and model CMS. That keeps any possible modeling error conservative and really impossible. As you'll see later, the truth may be as high as I mentioned, 97%, but there's no possible way it's lower than 85% and why I used that number is all the sources openly admit to that level.

FIGURE 1, PLEASE KEEP IN MIND WHAT I SAID ABOUT THE GLOBAL FOREST DATA.

Owner class/ land class	Region			
	U.S.	North	South	West
	Million acres			
All owners	766	176	244	346
Timber land	521	167	210	144
Reserved forest	74	7	4	63
Other forest	172	2	31	139
National Forest	145	12	13	120
Timber land	98	10	12	75
Reserved forest	27	1	1	24
Other forest	20	0	0	20
Other public	176	35	20	122
Timber land	63	29	15	19
Reserved forest	47	5	3	39
Other forest	67	0	2	65
Private corporate	147	29	65	53
Timber land	111	29	61	21
Reserved forest	0	-	0	0
Other forest	36	0	4	32
Private non-corporate	298	100	147	51
Timber land	249	99	121	28
Reserved forest	0	0	0	0
Other forest	48	1	25	22

Figure 1 Complements od U.S. Forest Resource Facts and Historical Data, FS-1035.

This is valid data on who owns and controls the forests within the United States. This is referenceable; it'll do for this point.

Compared to Figure 1, the rest of the world is in the same boat made of wood, but deeper in the maturity hole. Let me point out in Figure 1: 766 million acres are forested within the United States, 445 million acres in private/corporate ownership, 176 million acres are state owned. Therefore, 572 million acres are on the "soon-to-be harvested again" list. That doesn't account for the "other forest" category, of 172 million acres, but you can bet those trees are or have been marketable. Ninety percent of the United States' Forest as 693 million acres, has been and will again be marketable, or, if you prefer the less politically correct term like I do, has been and will be "logged" again. Everything on this list lacks maturity. National parks and monuments are not listed here.

Now, let's talk about this chart's data validity. The reserved forest of seventy-four million acres. Such a nice way to say it, "reserved forest." I call malarkey on that. Much of that reserved forest was logged and will be logged again because most of those acres are managed by state forestry agencies, the Bureau of Land Management (BLM) or the U.S. Forest Service (USFS is part of USDA), and, you guessed it, have been, and will eventually be logged again. Or they're in regions that don't produce trees very well or are difficult or impossible to access. Think about wilderness areas then think how they are part of our strategic reserve, that's right harvestable during war. And when you hike through them you will find old stumps from harvesting earlier. Here's the point. We must think about a tree's maturity first and then how many acres there are of them. It's now mooted to discuss how many acres we have because we know it's impeded, harvestable, and absent of mature trees. Almost every bit of it.

Laws that make logging on public land also force those agencies managing them into marketing trees, even if they don't want to. And having worked in some of those agencies, I know most of the people in them don't want to cut trees down for the sake of timber markets. They'd like to see them grow old and die naturally. But it isn't their decision, solely. The fifty-two million acres protected by law as national parks and national monuments are not on the chart. Guess why. **First**, they are not considered marketable. **Second**, some 70% of that protected land isn't forested, and what is forested was likely logged before becoming protected. Yeah, no kidding.

    Many countries have National Parks just not enough acreage in them to really help global sequestration. Oh, and I want to be clear about the BLM and Forest Service forestry programs. They both do an excellent job with the tools they have in managing our forests for us. Political agendas applied by corporations get in their way. In addition, they lack sequestration knowledge today but not for much longer. CMS provides a maturity tool that makes their efforts and lands more productive for their local communities, the USA, and the world. Literally, CMS can put them back in the forest by killing much of the litigation and political pressure they currently suffer with sequestration facts. CMS makes those forests even more renewable and productive than ever before. They have millions of mature acres that need thinning to improve maturity. Credit given, now, I've got to blast them a little more. Here is some really tainted data on tree ages.

    a.

FIGURE 2, ANOTHER ESSENTIAL FACT, AND DATA INCONVENIENCE...

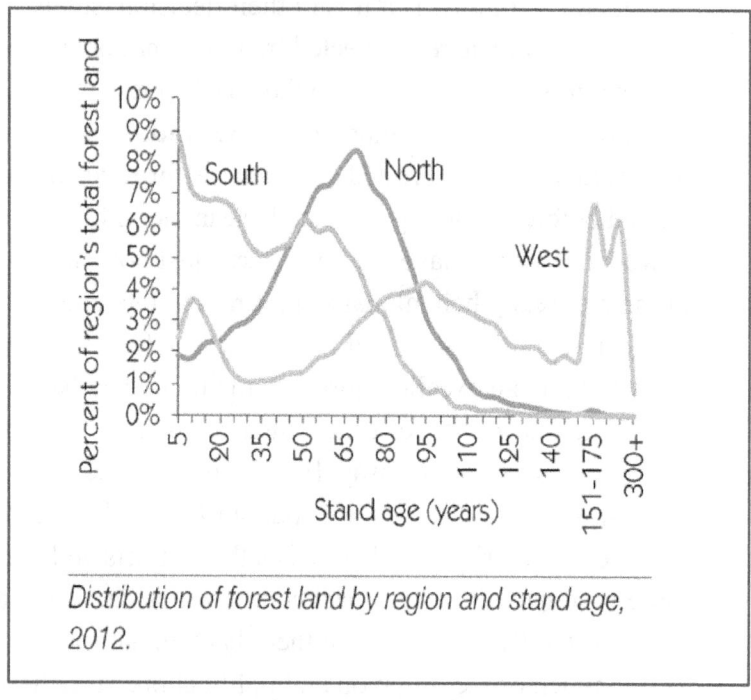

Distribution of forest land by region and stand age, 2012.

Figure 2 complements of: USDA Forest Service 2014, U.S. Forest Resource Facts and Historical Data, FS-1035

Forests stands over three hundred years of age only make up less than 1% of the United States' forests. LESS THAN 1%. The same can be said globally, except for the Amazon Forests'. Whereas 85-95% of the available global forest has been or will be logged again after regeneration. The less than 1% of old-growth are tucked-away two or three tree stands that are probably too difficult to log, considered seed trees for natural regeneration purposes, and for the most part sought after by timber companies. Or in a tourist destination, like a campground or hiking area where they are left alone.

Although, in those settings' vandalism is becoming more of a problem.

Now the first deception. A stand could be a couple of trees or tens of acres. This chart provides stands, not acres. So, the stand could be made up of mainly saplings with a few older trees sticking out and still counted as one hundred years old.

Figure 2 graphically shows an earlier point I made about data accuracy. Be careful looking at Figure 2. Figure 2 is entirely constructed by those with the intent of deceiving you, the public. Maybe not intentionally because as I mentioned, sequestration and maturity are not the point of this people pleasing graph. The statistical smoke and mirrors used is explained like this. Figure 2 is only scaled up to 10% and the age depicted starts at 5 years old. For example, if you look at where the "South" line begins, just below 9%, or any line, know this; most tree ages are not in the given scale, meaning they're either less than 5 years old or non-planted lands at the time of the survey. Unfortunately, the latter of which is the majority, as in mostly unplanted, and mostly in natural regeneration with a few trees poking up. Another look at Google maps satellite imagery provides more proof if you need it.

The mathematical evidence is Figure 2 is the graph did not begin at tree age 0 or 1. As a result it is not allowed higher than 10%. This resulted in nothing younger than 5 years old being depicted, AND there are certainly no non-planted clear-cuts (bare ground) influencing the stand-ages nor their percentage of all stands. That constrained the chart's age range to under 10%. And provided trees an increase in age shown. The suspicion of doing that intentionally becomes obvious. It makes the percentage of land filled artificially higher while it artificially increases tree ages provided. So, they left it out to their intended demographics perception's benefit. But was it on purpose? Possibly for the public's benefit? I can only wonder.

They do lack sequestration knowledge and thus the importance of tree and forest (stand) age correctly demonstrated.

That makes a point about available sequestration data, it's not part of the demographics that author(s) serves. This chart has been used and updated annually now for decades. Figure 2 is from 2014. Most years up to the 2024 version can be looked up for yourself just Google "FS-1035." But nothing has really changed, it's about pleasing perception and not criteria to directly support sequestration. I think they just lacked the data, and that could be true. Seriously, what else could it be? The data they receive from private corporations is what they used, so it's likely they didn't do the survey or expand the criteria to represent reality. But this chart, like many others, really begs to be reconciled with exact data on tree maturity and land usage instead of its attempt at feeling good for its demographics, doesn't it?

Granted, CMS's point on maturity still sticks out like a sore thumb even with Figure 2's not so stellar, but referenceable data set. Finding a 100-plus-year-old tree, here or globally (outside the Amazon or any national park or monument) is like finding a matching sock in the dark with twelve different colors available. If there are any elsewhere, I certainly don't know where they'd be on any scale. There are some old and mature growths on government managed lands here in the United States. I've seen BLM and USFS lands with old-growth trees here in Oregon, but it's rare to find more than a handful of them together. And they are always surrounded by immature stands. The BLM has most of them locked up behind gates as seed trees. The Forest Service hardly has any in comparison. Even combining Forest Service and BLM lands there's not enough of them to fill much more than a tiny national park in all of the United States managed timber stands.

What those government lands on the west coast do have is a good jump on maturity. They have a lot of thirty-year old plus stands thanks to the Spotted Owl becoming endangered in

1990. Hooty becoming listed as an endangered species made it difficult for government agencies to harvest trees. But harvesting is possible, it's still happening and on a regular basis. The harvesting method they conduct today is mostly thinning, however it is not *natural attrition harvesting*. Not anywhere near it. Instead, they use *demand driven forestry* thinning because they cut older and better growing trees with thinning's disguised as environmental or habitat restoration and not to advance maturity.

Today, government managed maturing replanted stands all await *demand driven forestry* harvesting, deception based thinning improvements, and other efforts that don't promote maturity. At least they get a good job for not clear cutting anymore, I hope. Their trying to bring that back. Anyway, time will tell if they listened.

Overseas and in the remaining America's, to include Canada they're lucky if they have 30-year-old regenerated trees and replants.

Now remember what I said about defining an old-growth, Figure 2 certainly makes the governments contradiction in terms relevant to proving their deception. 1% in Figure 2 of 145 million acres in Figure 1, page 97, isn't 80 million acres of old-growth they now are trying to claim with their new definition. 1% would only be 1.45 million acres. But according to both figures that doesn't check either. Their deception is defined by their own admission, yep, no doubt their hand is in the cookie jar and caught.

A SAD BUT TRUE REALITY

Consider the devastating fire in Paris's Notre-Dame Cathedral. The fire destroyed much of the historic structure. While the decision to rebuild it was swift, it came at a significant environmental cost. Plus, this makes a point about societies less than benevolent view on forestry use.

Instead of modernizing with steel, they chose historic restoration using wood; specifically, very mature, and almost old-growth oak. The trees were growing between 140-160 years, they were critical for atmospheric sequestration. Unfortunately, their huge size also made them perfect for replacement timber supports.

Those mature oaks were among the best for CO_2 sequestration, and their removal has lasting consequences. Cutting them down was a grave mistake that exacerbated everyone's climate crisis. Those trees had played a vital role in regulating the climate, and their loss is deeply felt and not easily recovered.

The restorers had to scour the world to find those mature oaks, ultimately using the Vatican's own reserve when no others could be located. In doing so, they unknowingly created an impact that will persist for the next 140-160 years.

The rebuilding of Notre-Dame required 1,400 plus mature oak trees. The environmental impact is not just immediate but stretches far into the future. By choosing to rebuild with wood, France increased its carbon footprint, contributing to emissions and methane in their microclimate. This impact includes emissions from the harvested area and wood waste.

Consider the *carbon hump* created by this decision. Cutting down those trees adds over 1.5 million lbs. of CO_2 to the local atmosphere in the first year alone, and this amount continues each subsequent year until the trees' sequestration is substituted with identical maturity. Moreover, this action has increased France's and Earth's *carbon hump*, contributing additional forest harvesting emissions and methane to atmosphere; likely close to a million pounds per year for up to thirty years as the harvested area, wood product waste, and construction scraps gradually decompose or are burned.

Notre-Dame's restoration, while visually stunning, has come at the cost of significant environmental damage. The notion that these trees were renewable or carbon-neutral is misleading. Up to 70% of a harvested tree is waste and not used unless the sawmill burns it to create electricity to run their mill. Thus, converting it into emissions that increase the hump. Today, thanks to advancements in sequestration science, we understand the true cost of such actions. We now understand the thousands of tonnes of carbon stored in Notre-Dame's timber construction is not comparable to the millions of tonnes of CO_2 now being added to the atmosphere because of the now absent sequestration and *carbon hump* components.

An easy decision was made while under the influence of wood products renewability's half-truth. That is exactly the kind of decision that is now proven to have created and is accelerating global warming. Decision processes we must stop telling each other to use.

I do hope this example serves as a reminder of maturities environmental impact. An impact prone to affect generations to come, just as we have been affected by our predecessors. To me now, Notre-Dame's restoration really wasn't worth it. Now restored, it stands proudly. A testament to human sacred beliefs. An icon that marks the path towards human extinction.

France, you are not to blame. You did not know of the CMS study, but now you do.

SO, HERE IS THE PROBLEM, AND IT HAS ALWAYS BEEN HERE.

First, societies' economic rules and forestry demand keep trees from maturing, not just a few of them, almost all of them. **Second**, yes, there are an estimated three trillion trees growing globally. Yes, it's good we have a lot of trees, but again, the majority of them are incredibly immature and also slated to be harvested incorrectly as a crop of corn. **Third**, those immature trees everywhere are also firmly anchored in the *carbon hump*. In fact, just before they emerge from the *carbon hump* they will be harvested and that rebuilds the *hump* and significantly adds to global warming. **Fourth**, the saplings we see today also must survive a very natural attrition rate of one maturing for every four or five saplings currently growing. So, three trillion immature trees really aren't that many trees at all. It's more like 450 billion mature trees (not old-growth) surrounded by 2.8 trillion immature trees still in the *carbon hump*. **Fifth**, only 450 billion of those 2.8 trillion will appear from the hump and that takes over thirty years of time. When they do they get harvested! **Sixth**, all the while, more immature trees sprout up to take the place of the fallen. And they do nothing to stop global warming. They are caught within the *demand driven forestry* cycle. Totally depressing but true. **Seventh**, Earth is in a *carbon hump,* so all trees mature or not suffer *accumulating decline* and share the human endangerment in becoming extinct. Humans will follow plants into the abyss.

In today's forestry management, saplings, replants, and natural regeneration can't and won't do anything to abate global warming without more maturity. Over eons, tree immaturity has become the larger element in the outcome of climate changing conditions being geoengineered. The *carbon hump*, harvest rotations, and the attrition rate guarantee this modernized problem by not allowing maturity to increase whatsoever, it is in decline. The current forestry scheme only results in *impeded fast-cycle CO_2 sink*s being perpetuated with industrial efficacy and at a globalized scale.

REPLANTING IS GOOD, MOSTLY.

If you're into helping environments by replanting trees, you'll need to step it up by doing five to ten times more in your afforestation efforts to make a serious global warming impact many decades from now. Also, try to focus on restoring forest lands and not the replanting of some timber company's harvest years from now, which is happening a lot these days with some nefarious nonprofits. And keep in mind, you're playing the long game with replanting. Afforestation efforts cannot stop global warming today like increasing tree maturity can, but afforestation activities today help abate climate changing conditions in the future.

PLAYING THE LONG GAME IS WELL WORTH THE EFFORT.

To repeat, maturing trees that made it past the *carbon hump* is a faster acting cure that is needed now. Continue to replant but make protecting old-growth and maturing trees we already have the priority.

Here's why increasing maturity is a better idea today than replanting for the future. It's a quicker atmospheric and economic return on our investment. Accumulating a few billion global trees that appeared from the *carbon hump* over 15-20 years can occupy up to 333 million acres of existing forest land and make it work like it's supposed to. As you'll see in the next two sections, that makes a serious impact on global warming from day one. As a bonus, doing so eventually repays for itself.

In contrast, replanting the 30% of forestry land that has disappeared, around 4 billion acres, costs a lot more, displaces millions from their homes, and significantly decreases agricultural production. And then there is this, it won't do anything to cure climate change for many decades. Unless we can replant hundreds of trillions of trees. Which we can't. We must use what we have available to beat global warming in time, and that's forestry acres past their *carbon hump*. But please keep replanting everywhere we can! I know I'm going to keep playing the replanting long game every chance I'm afforded. Arbor day was made for me!

Reckoning Tree Data

Now consider this quote taken from Our world in Data, July of 2024: *"a study published in the journal* Nature, *Thomas Crowther, and colleagues (2015) mapped tree density across the world. To do this, they utilized 429,775 ground-sourced measurements of tree density from every continent on Earth to generate a global map of forest trees. They define a tree as a plant with woody stems larger than 10 cm in diameter at DBH (DBH, diameter at breast height), (10 cm equals 3.93 inches.)*

They estimated there were approximately 3.04 trillion trees in the world. The authors also estimated that over 15 billion trees are cut down each year, and the global number of trees has fallen by almost half (46%) since the start of human civilization."

That's around half a trillion trees harvested every 30th year by Crowther's OVER 15 billion harvested annually estimate. If they make it for a thirty-year period, IE, if they become 30 years old, they got lucky. Add natural attrition, forest fires, and all the trees succumbing to *accumulating declines* these days and then remember that OVER 15 billion are harvested. And the ones on unprotected acres, 9.8 billion acres out of 10.3. Well, I think Crowther's 15 billion harvested is too conservative and does not reflect other tree demises like fire and bugs. Not with all of today's forestry carnage brought about by immaturity could they. What may be a better way to say it than Crowther did is the age of those three trillion trees are reset back to zero before they reach 60 years old.

Knowing that reset happens, CMS estimated 30 billion trees do appear from their *carbon hump* every year. These trees can be used in CMS mitigation plans to end global warming. That number seems reasonable if not ultra conservative. So, I'd like to believe adding 3 billion trees or more annually to CMS's mitigation effort is at the least practical. Well, it's at least possible.

Since we're working with 9.88 billion unprotected forest acres globally, out of 10.3 billion (4 billion hectares) using CMS on 334 million acres over 20 years also seems very reasonable, I'm sure it's not that easy. But then again, we aren't stopping the use of forestry, were making it more efficient while making it compliant to sequestration value. You'd think all that should help avoid problems. CMS has other ways planned; but let's stick with this way because it is practical and not overreaching, by much.

SEQUESTRATION ACCUMULATION EFFECT, MITIGATION.

Let's graph CMS's 1st goal and its planned global experiment in implementation. As to details, those trees that beat the hump immediately begin sequestering on average around 190 pounds of CO_2 annually from the atmosphere and naturally increase every year from there on. The emissions from their surroundings and products made have finally dried up. They add to the annually increasing effect of CMS's 1st mitigation goal as it applies, the **Sequestration Accumulation Effect**. This demonstration is a perfect world scenario but accurate in its assessment. There are variables that may affect the overall outcome, which is why I've included a margin of error annually as vertical bars on the graph's annual points.

FIGURE 3, CARBON HUMP EMERGENCE MITIGATION. THE YEARS USED IN THE PLAN ARE THE BOTTOM, X AXIS, AND ANNUAL GIGATONNES CO_2 ARE ALONG THE VERTICAL, Y AXIS.

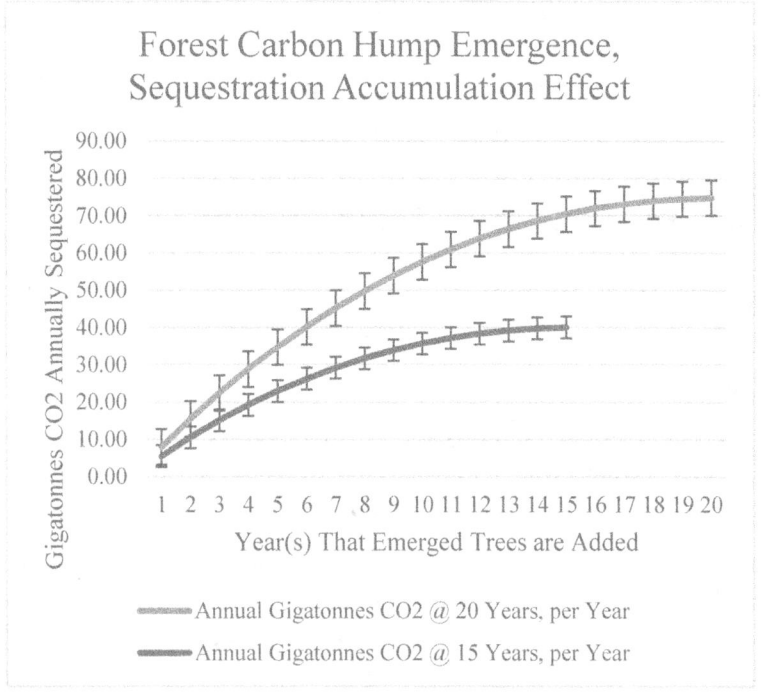

Figure 3, the Sequestration Accumulation Effect happens naturally as trees emerge from their *carbon hump* and mature. Figure 3 uses 3 billion trees added annually and typical tree growth models for sequestration rates. Annual sequestration rates are graphed to prove maturities sequestration increase over an annual period during either a 15 or 20-year duration. For example, year twenty of the 20 years line is equal to 20-year duration of the first 3 billion trees plus the annual addition of 3 billion more trees and the years and sequestration they accumulated. So, it is the accumulating sequestration growth added to the annual amount of the other periods restricted by their participation in time allowed. That's a mouthful,

Basically, it is a positive feedback loop tree growth provides added to the number of trees and totaled up each year. The tree growth levels that maturity provide increase sequestration and add to the tree accumulation that also increase each year. Okay, there is one more way to explain it.

For example, in the first year, the first three billion trees will remove somewhere around 0.28 gigatonnes CO_2 from the atmosphere. All by themselves. Initially, that isn't very much. But the compounding effect from three billion trees allowed us to keep growing while adding three billion more each year that does make a significant difference quickly.

With this method and on the 15th year, we're annually removing **41 gigatonnes** of CO_2 from our atmosphere (or a lot more given species and region used) and the positive feedback continues increasing that annual amount yearly, even if we stop adding trees. If we don't, at the 20th year, by adding five more years the effort is removing **75 gigatonnes** of CO_2 annually from our atmosphere. Note: These are clean gigatonnes and not in a *carbon hump*. Well, technically their still in the global *carbon hump* like our remaining old-growth but we did decrease that hump by **516 gigatonnes** over 20 years. These results are in addition to today's global sequestration levels and improving them each growth cycle.

This outcome improves today's paltry and increasingly failing global forestry sequestration levels, roughly estimated between **92-96 gigatonnes** annually, after the hump takes its bite. So, in 20-years the accumulating effect removed **516 gigatonnes** CO_2 more than what's happening today, or really in the last 200 years. All that occurred by increasing forest maturity.

This method ends up using sixty billion trees of the almost 3 trillion available (2 percent of all existing trees). Add that the method only used around 3.2 percent of our global forestry acreage, 333,333,333 acres. With maturity's effect on sequestration, it really doesn't take much to make a huge dent in global warming.

I did say mature sequestration is very powerful stuff and I hope this basic exercise demonstrated just how powerful it is. But is it enough? No, not really. Not all by itself. We need to move faster, and we can.

Unfortunately, using 3.2% of acreage available is not enough to cure global warming in 15-20 years. It comes very close but misses the 25-35-year deadline. We showed earlier on page 87 Earth was roughly 22gt from an emission to sequestration annual balance a few years ago. But if we want to cure global warming the ridiculous amount of CO_2 stacked into Earth's atmosphere must be addressed.

Around **1,541 gigatonnes** of CO_2 need's sequestered from Earth's atmosphere. That amount, plus the balance is needed to end global warming's reign. And that is a lot. The good news, this method starts pulling from atmospheric residence 8 years after it obtains the annual emissions to sequestration balance. Unfortunately, it requires something like 45 years to pull it all out. That's if human emission levels don't increase and they will. Since we don't have 45 years, we have 25 maybe 35 at best. And, to avoid any unforeseen problems from *accumulating declines*. We need to scale-up. Each growth cycle we do, that 25-35 years becomes 26-36 then 27-37 and so on. From July 2024.

That explained, we can actually cure global warming within 15 to 20 years with this method scaled up. In 5-10 years, we can be reeling it in measurably. Plus, the effort is continuing to escalate its results every year after; but again, it is not enough to fix it in time, not without scaling up.

The math provided to scale up is easy. Double the effort double the results. To what end you might ask. To where we need to be globally, I will answer that after explaining some details needed and later in this book. For now, if we desire a quicker fix that's required, we must increase the effort. The nice thing is we can because all CMS mitigation is *proportional*. Not in just this effort either. Nature is kind like that. The more land and maturing trees the less time needed to end global warming. After CMS mitigation is implemented at scale it gets even better than this example. *Proportionality* explains further but CMS's efforts don't stop with only this scenario. For example, the more mature the trees are the less acreage needed, and less trees are required. Which is a trick we'll apply later. For now, if only it could be this easy, it can't and won't. There are details requiring attention.

One problem to address, and likely to turn into a brawl is simple to relate. Those trees that make it past the hump are on any given day in a harvest rotation. We'll have to compete for them, and they are not free. Right now, we can't stop or even slow the harvest down. Which also makes all this idealistic and not reality, at least not yet. As I've mentioned in a different way, reality today is a combination of human behavior that doesn't want to change its blundering into the abyss. But those who are willing to change the world one tree at a time are out there by the billions. Which is exactly what this plan needs.

REMAINING OLD-GROWTH.

Think about this CMS fact on global old and mature growth: Roughly 15% remains of mature global forest, sort of. That forest remaining is estimated to be 750 million to 1.545 billion acres out of 10.3 billion total forest acres. The United Nations says 1.5 billion, I say 750 million. I hope I can explain 1.545 billion is not likely, or really provable. The 15% of forest land the U.N. talks about as protected is and has been logged and most of it is still in a dictators portfolio, in a corrupt or weak government, or being tapped by local demands for resources. Basically, a lot of it is in a harvest rotation, wood fuel is being used beyond its regenerating ability or it provides economic means. Or it is in a region that is highly unproductive for trees.

In short, that 15% looks good on United Nations reports but consists of regions and species that might be dwarf trees, has no trees at all, was logged before becoming protected, or is under some kind of *demand driven forestry*. Technically, I estimate mature tree sequestration on productive lands occupies under one million acres to roughly half of the U.N.'s estimated 1.5 billion acres of protected land, 750 million acres. That land encompasses something like 120 billion mature and old-growth trees out of 3 trillion trees. A miniscule portion, need, I say more?

Some of those forest lands do hold maturity, the only thing saving humans from a complete climate collapse. But there is no such thing as a perfect forest, not anymore. The millions of protected acres I've researched don't have old-growth, maturity, or are not in tree growing regions. Parts in tree growing regions can even be under natural regeneration due to harvesting before becoming protected. Most global protection mainly happened during the 1990's-2010 and a little since and a little too late. Now add human impacts like settlements and logging in the Amazon, naturally occurring and human-caused fires, and other *accumulating declines* like the ongoing global drought and it's just not near enough.

It's really only worth mentioning the Amazon forests and longer standing national parks. The old-growth those land's contain provides the only real still intact sequestration mechanisms that fight against our climate accelerating out of control right now. That's not totally true because of the *carbon hump* realities deconstructing their efforts otherwise-we are in a climate collapse. They are all surrounded by CO_2 releasing harvest sites whose *carbon hump* passes onto them.

For the record, we need up to 30% of global trees over the *carbon hump* and maturing to get out of global warming's grip the fastest way possible.

Now allow me to tie this altogether with a typical example. Yellowstone National Park. Yellowstone is not a good tree growing region, it's a temperate ecosystem. The tree species there are not known to grow fast, nor become really big and tall like rainforests or coastal regions. Some trees do but you won't find road signs in Yellowstone directing you to huge trees like Sequoia, Crater Lake, or Redwood's national parks. In addition, recovering from the fires takes time. Yellowstone burns frequently. The 1988 Yellowstone fire took out it's maturity and that maturity has yet to recover. The fire also added a fire induced *carbon hump*, but it is looking better every day as it slowly regrows! Yellowstone is 2,212,520 acres and in a somewhat unproductive region plus, humans. Every year, over 3,000,000 visitor's and their vehicles add to the 1988 fire's emissions and guarantee it's *carbon hump*. The point is all factors must be accounted for to determine a forest's or regions sequestration ability. You just can't say it's protected, and it magically obtains *sequestration value*.

    Let's add to the example, the Amazon region has hundreds of millions of acres being logged or in regeneration at any time. Sure, the Amazon Forest area started out as 1.7 billion acres and spans six countries. However, by the UN's own admission 26%, 442 million acres was lost to land use modification by 2022 leaving only 1.26 billion acres and is still decreasing, daily. In fact, 30 million acres have burned this year as I write. So, many of those 1.7 billion acres are gone and the chainsaws, fires, and bulldozers are still working against maturity. Which makes another point.

    What I find most ironic about the Amazon is the conservation efforts the rest of the world forces on the people of those six countries. We force them to not use their forestry resources. That makes them practice illegal forestry without a care in managing it because and as the norm they get away with it and then have no accountability in taking care of it. To me our double standard makes it all worse. Starting with it was fine and dandy to do it to our own forests, but they can't? A double standard made by late arriving science I suppose. But the truth is more likely monopoly driven. Making it legal and allowing them to become stewards seems the better practice. I'll state for the record again, it is not using forestry resources that are bad, it's a lack of stewardship. That is, the stewardship that discounts maturity is bad.

Natural attrition harvesting is the answer for the Amazon forests and its people. It is for our own forestry management debaucheries as well. That is the answer sequestration science has provided and the only correct one.

Time is of the essence. Knowing what's needed to reverse global warming concerns me deeply; knowing we still have some sequestration to work with provides some comfort. We can take some refuge in knowing Earth still has a limited supply of measurable *fast-cycle CO_2 sink* ability. Remaining old-growth provides it. As a bonus a lot of sequestration also occurs, away from the Amazon Rainforests and global national parks. Crops, grasslands, and older trees beating the hump everywhere are contributing. Even algae and some bacteria contribute. Even trees like the ones by the county courthouse, in someone's yard, and within your local parks do. The few remaining old-growth on federally managed lands and the maturity Hooty derived is also beginning to really help. Old-growth's horizontal forests of limbs are out there, just not enough of them, yet.

WHAT ARE THE ODDS

The current state of Earth's global forestry sinks offers limited odds for making it much longer as Earth's oblivious to accountability guests. Eviction from our home is becoming eminent. Our survival here is now a bad bet even for the riskiest of climate denying gamblers. Winning against global warming now requires more skill than luck. Our luck is used up but it got us to here. Sure, CMS skills can make more forestry luck as we blunder on, but will it be made quickly enough to win the bet we've all pitched in on and made against Mother Nature?

The fastest cure comes from forest acres over their *carbon hump*. That limits the available supply because those trees are coveted by industry. To get enough, we have to go global right away. The older the trees, the better. Surprisingly, the effect on our atmosphere happens rapidly. But the local effect, well, that effect is even more amazing by creating strong microclimates within our global climates. Which is partially due to CO_2 atmospheric residence conditions. So, one country won't care what another country is using to produce energy for its people? No, not really. There are no free rides in sequestration. Resident conditions are why.

ATMOSPHERIC RESIDENCE CONDITIONS GLOBALLY APPLIED.

Okay, the influence of CO_2 residence conditions on globalized climates can be summarized by example. Say Europe's emissions go into the atmosphere of Europe. A given right? Well, not really because the world spins and the winds blow, weather happens, and sequestration is paltry and even nonexistent in areas where people live. All of which influences where those emissions end up. Its randomized chaos.

Because some countries don't have much in the way of mature forests capable of sequestering their emissions, CO_2 is going to hang around but eventually move somewhere else via the influences I mentioned. So, Europe's CO_2 emissions eventually end up in the Americas or some other part of the world that can sequester them. Or used to sequester them. But that could take an uncalculatable time. Perhaps a year or a decade to happen under good sequestration conditions. Under today's paltry sequestration conditions eons at best but right now, probably never. And since humans aren't going to make it that long without CMS its moot. However, the time taken between emissions to sequestered can be measured in residence time. So, the time or duration emitted CO_2 is within the atmosphere is considered as CO_2 atmospheric residence time.

Residence time is accounted for by CMS as increasing *proportionally* to sequestration decreasing. The amount of CO_2 emissions that enter the atmosphere has no effect on residence today. But you guessed it, it used to.

CO_2 emissions should be proportional to residence time, or you can say they should be in some balance. As in, the more CO_2 emissions in the atmosphere the less time it takes for healthy and mature photosynthesis to absorb them out. Now that CO_2 emissions have nowhere else to go but into residence conditions, our atmosphere is becoming saturated. Where the actual point of saturation is, I don't know. But I do know what happens if we hit CO_2 atmospheric saturation. Extinction. The reason we know its extinction is it's all happened before on Earth, and more than once. Seven different times Earth has put the ax to creatures who became fossilized records. Technically six times on its own and one big meteor that got the dinosaurs. We're next if we don't do something about global warming. A more positive part filled with hope! Humans are the only creatures pending an extinction who are capable of doing something about it. That is supposed to provide hope, I hope.

A PEEK AT MORE RESULTS ANYONE? HOPE TO EMBRACE.

Respectfully, a large-enough 30-year-old forest that's allowed to age another 20 years can significantly impact climate change without much ado. Get enough CMS managed forests to become 50-80 years old and CMS stewardship can absorb all human and natural CO_2 emissions and eliminate excessive atmospheric levels.

So how large is large? Well, I mentioned acquiring 2% of global trees and 3.2% of forestry acres over 20 years in the Carbon Hump Emergence plan, but truthfully to do it quicker and be assured of results we'll need more of the 9.88 **billion** acres of not protected forestry available (of the 10.3 billion forested acres). Why? Because nothing goes according to plan, and we really don't have time to mess around to find out. To fight Murphy's law, we need to be conservative by increasing our numbers in order to guarantee results. And add one critical warning.

First and without delay, protect our remaining old-growth and mature trees, which currently and cleanly sequester approximately **92-96 gigatonnes** of CO_2 annually, after *carbon hump* deductions. That should be hundreds of **gigatonnes** without *carbon hump* deductions made with current forestry harvesting. Which is a very good reason to make it all work better with *natural attrition harvesting* globally. So, old-growth COULD and SHOULD occupy around 3.9 billion acres globally, and eventually can with CMS maturity restoration adding to it over time. So, protect what we have first then start buying acres and acres of younger trees emerging from their localized *carbon hump* (20-30 years old and up). Don't cut them down, or else! Thin out the dead and dying for the biomass and let the healthy trees branch out and grow bigger! That also helps dissolve the old-growth *carbon hump* one tree at a time. That also creates an explosion of sorts. A compounding effect happens that promises huge decreases in atmospheric ppm levels. And yes, emission reductions help a little to remove that hump and why they are worth doing.

Keep in mind, trees we add might require thinning upfront and then every few decades because of natural reseeding and attrition, but that's okay so long as it's done with *natural attrition harvesting* criteria. It's better for the sequestration and maturity of that forest and Earth's *carbon hump*!

Now, how much forest can we get? Well, we can get more forests with money and none without money. Reality is what it is. Now looking around the global forest for-sale list says plenty. As in, CMS is very doable, for a price and it won't happen without funds. With, we can apply *proportionality* tricks to help end global warming.

One trick goes like this; *proportionality* works for us. Immature trees require more acres while old-growth or mature trees need fewer acres and fewer trees to increase *sequestration value*. With that trick our goal is to achieve a balance between the two, while more immature trees and even some afforestation is added (where it can be). All that eventually boils into a few postage-stamp-sized tree plots on the toy Earth globe in your childhood classroom. That's all it'll take to end global warming from running Earth out from underneath us. We know this because that is what humans had prior to 1700 and although that wasn't optimum, it's a good goal that improves each year maturity bloom's. But it'll take time to get there. Trees grow slower and much longer than humans. But proper tree maturity in those postage stamps on the toy globe will negate current and past emissions that make global warming the problem it is.

My inner tree hugger hates to admit to this, but that plan is also around 92% more efficient at growing biomass. Whoa, you mean 92% more biomass in 10-15 years? No, I mean up to 92 times more biomass production in 20 years because of that 3-15% annual growth rate of trees improving as the atmospheric ppm level drops and starts to end our global drought. It all seems worth it, doesn't it?

For now, CMS showed us sequestration is powerful stuff and can change our luck with skill. Thanks to tree growth rates and *proportionality*. Unfortunately, humans have limited Earth's sequestration ability, so curing climate change will require many acres initially, but not as many years as you'd think are needed to make a very noticeable difference. More acres are faster with less localized *carbon hump* to deal with. More old-growth maturing means fewer acres and bigger atmospheric impact. All tricks CMS pointed out that can end global warming. And we know it works because it still does and had for hundreds of millions of years before humans collectively messed it up.

DEEP DIVE INTO THE DATA POOL

THE MASS TIMBER PROGRAM

The Mass Timber program makes it possible for smaller and smaller trees to become marketable as peelers and chips. That makes it a demand *driven forestry practice*. However, technically, mass timber also uses biomass more efficiently by adding adhesives, pressure, and heat. Mass timber products are glued together to make wood products of larger dimensions. The thing is, products derived from mature tree growth made all the Mass Timber products naturally, without the glue or extra processing but the old ways wasted up to 60-70% of the trees biomass. Mass Timber can use that waste and therefore has increased biomass efficiency that equates into less *carbon hump*. So, CMS's increases in biomass efficiency can actually produce those products while ending global warming but that also produces the 60-70% waste component. Supplementally, Mass Timber can use the thinning's from CMS to make its products while ending global warming. It's respectable progress so long as Mass Timber promotes maturity first.

Therefore, CMS and I don't have any real complaints about the current well-funded government and academic initiative named Mass Timber. It is an advancement in human self-domestication, a tool. Again, so long as it promotes maturity first in harvesting.

So, the Mass Timber program has benefits in forestry efficiency CMS likes. They can use more of the dead and dying tree found in thinning's than conventional practices do. While avoiding a lot of waste, which should leave more trees in the forest after answering human demand, under *natural attrition harvesting*. Therefore, support *natural attrition harvesting* Mass Timber, and you'll have my full support.

Don't do that and you're damning all of us, to include yourself. Not a difficult choice, is it?

THE CMS CLIMATE CHANGING DATUMS.

CMS's first datum, the 1850ish datum, is just prior to the worst part of *constrained deforestation* accelerating. With that acceleration our atmospheric CO_2 ppm levels also greatly accelerated. Prior to 1850ish we still had enough maturity to do something about it but after and since, not so much. In comparison, the 1950ish datum is a real eye popper.

CMS calls those two marks in time its *climate change datums*. The first point in time we should not be proud of is 1850ish. That datum formed like its counterpart, a result of *constrained deforestation* advancing across the globe and that resulting in a correlation with CO_2 ppm levels increasing during the timeline. The 1950ish datum is a continuation of 1850ish but achieves significance as a pinnacle. It can also be interpreted as Mother Nature's final warning.

The datums' sources is NOAA. Pre-and-post 1958 ppm data was collected from ice core samples and then flask measurements from 1958 on at Maunu Loa Observatory.

Please keep in mind, the 1850ish nor the 1950ish datum are not when climate change started, they are years Earth's atmosphere and sequestration correlated from accumulated events that resulted in noteworthy atmospheric changes.

Around the **1850ish datum**, millions of years of Earth's sequestration and emissions balance degraded further. During this period, the datum shows CO_2 began accumulating in atmospheric residence longer and began changing Earth's climate faster. What can be said accurately about 1850ish and why I chose it? The certainty comes from the correlation between historic forestry demand and atmospheric CO_2 ppm both taking off from norms proven previously. Those two things correlate because of what the study calls "*constrained* *deforestation"* which began spreading around the globe like a slow-moving pandemic around 1700 on. Mostly by wooden ship building enterprises and then wood used as fuel to produce steam and charcoal. And it really showed up in the ppm data as railroads spread it year after year. It continues to plague us as it continues to spread even to this day.

I know, I'm avoiding the 1950ish datum, it is painful to communicate. I promise to explain 1950ish in a couple of chapters. So, lets explain *constrained deforestation* before going further. Its in the glossary, but it's also important.

Constrained Deforestation.

CMS's *"constrained deforestation"* is forestry that is physically kept and not allowed to regrow into a forestry normal. *Constrained deforestation* is opposed to the typical definition of "forest degradation," where forestry is at some point expected to regrow into a forestry normal. That term seems to imply renewability, my term does not, and for good reason. The action is constrained by the destructive circumstances perpetuating it's results. Basically, it only ends when humans decide to not do it anymore.

The constrained effect is typically due to human interference supporting inefficiencies like but not limited to, low duration harvest timelines (immaturity), or unintended but consequential forest impediments brought on by weather, fires, climate changes, biological events, basically all *accumulating declines* known. Eventually it leads to one step further, *unconstrained deforestation*. Which is when forests land becomes something other than a forest permanently. The result of *accumulating declines* winning, or human impacts. Thus, becoming unconstrained in efforts to end forests and keep them from returning.

Almost everything that could be easily logged was logged as *convenient forestry* by 1850. But *demand driven forestry* practices continued to expand globally. That is *constrained deforestation* because all global *convenient forestry* has yet to recover a forest normal, its maturity. Humans keep that from happening. That is a reoccurring problem in all *convenient forests* even before 1500 in Europe and Asia. But by 1850ish it was happening in the Americas and across the Pacific. Today, it has all been logged many times or was cleared for land use modifications like urban development, farming, mining, or manufacturing. Just after 1850, the harvesting of the Amazon River basin and the coasts of the Americas significantly increased. In short, humans entered many of the last old-growth stands the world had to offer. And most of the east coast of the Americas *convenient forestry* was in short duration harvest rotations, used up, or gone by 1850.

Populations increased; more laborers all over the world were becoming economically happy to cut old-growth down in exchange for a few coins. They didn't know any better and needed to survive; so, they went logging. It was around 1990-2000 that the USA and most of the America's had harvested all but the protected forests or stands too remote to profit from. Access improvements with transportation advancements and the human population expanding its labor pool made it all happen quickly.

Railroad construction made *demand driven forestry's* expansion possible and that made *constrained deforestation* entrench into forestry management plans. Railroads needed unbelievable numbers of rail cross timbers (railroad ties), lumber to build new towns and cities, and firewood for their steam-driven equipment. Inevitably, rail lines began to extend their reach into the vast old-growth forests of the Americas which brought those forests into marketable status. Many rail lines were constructed for the sole purpose of harvesting old growth from remote areas.

The world's great railroad expansion used every stick of wood it could get. I'm not kidding. Almost all of the *convenient forestry* globally, even Africa's vastness could not save it from railroad access. Everything steam-powered before 1850 used wood as a fuel until coal replaced it at a higher cost. Coal only replaced wood around 1890 because steam power ran out of *convenient forestry* from which to get the much better and way more useful wood fuel cheaply. That decision was forced on humans. And if wood is available, they still use it for cooking, heat, and steam in many places around the globe! In fact, most modern wood producers still use it as fuel to generate their electrical power. What I must say to that is this. Thanks to the powers that be that we didn't strip the Earth of the harder to access old-growth trees back then, or else climate change would have already made us extinct. We dodged a bullet, all thanks to coal and other fossil fuels! Ahh, did I say that out loud? Yeah, I did.

It was a good thing for sure whether I like fossil fuels or not. Energy derived from firewood could not support the populations needs then or now. There simply is not enough forestry to produce enough wood. Literally, then, and more so now; Humans could burn all global forestry and be left in the dark within a few years. In Africa, the use of wood as a cooking and heating fuel has decimating many of their forests. If there are a people who deserve a charitable natural gas line you'll find them in Africa's forested regions.

By 1915 CE, even the remotest global forests were being made marketable by railroad/steam expansion. Around then, gasoline-powered trucks also started logging and extended humans further into old-growth and virgin forests on the America's west coasts. Since then, very few places on Earth growing trees are *inconvenient forestry*. The Amazon Forests are likely the last *inconvenient forests* on Earth.

Today, we've stepped it up. Trucks didn't have the up or downhill restrictions of railroads or the expense, so they quickly became preferred in the 1930's and still are. They went where railroads couldn't. The much higher impact on forestry was felt by the forests as new roads were constructed. It was here that demand increased with a growing population's labor pool and world wars. Then something really hard on forests occurred. Quite possibly, the worst thing ever.

The early 1900s is also where the timber barons joined together with the U.S. government and promoted the first timber "conservation" program, called "nominal measurement." This program was not conservative. It was just the opposite. And not to be confused with what's happening today with the Government funded Mass Timber programs. Which is the same profit driven concept but different in efficacy to make products. Mass Timber lacks the horrendous environmental impact of nominal measurement. If it promotes maturity first.

LUMBER'S NOMINAL MEASUREMENT SYSTEM, ARTIFICIAL DEMAND.

The modern practice of lumber measurement uses a nominal measurement scale. Nominal measurement can be referred to as greed, bad forestry management, or what it really is: answering the public's demand for forestry products with a *constrained deforestation* answer. It is *demand driven forestry* that creates artificial demand and mega profits. We humans can sometimes be real suckers when we're told we're helping. I equate the nominal measurement scheme as the first environmental fleecing of the public. Unfortunately, fleecing schemes have since been increased and use nominal measurement's emotional architecture in propaganda. Propaganda designed to take advantage of the public's concern for the environment. Sound familiar, like emission reductions.

CMS predicts and warns like emission predictions just can't and don't accurately. CMS is with facts. So, all the guessing and taking advantage of our good intentions is over with. The tombstone epitaph for the emission reduction grifters should read, "It's a fact, I died because I did not live by global warmings facts." That is of course, if there's any of us left to engrave it.

Nominal measurement created massive artificial forestry demand. As in, one measured 2" x 4" lumber component (using load-calculation determinants) is equal to 1.25 to 1.5 multiples of the modern nominally measured 2 x 4 that actually measures as 1.5" x 3.5". That simply means to do the same work (as force or Newton's) you need not just one nominally measured 2 x 4, you have to buy two of them. And it affects all lumber products not just two by fours.

Nominal measurement also perpetuates fast cycle CO_2 impedance by restricting forestry recovery durations within ALL global forests used for contemporary lumber production. In gist, buying one 2 x 4 that measures 2" by 4" instead of 2 nominally measured boards is just better for everyone and everything except for the wads of cash in the timber barons' pockets. Conservation it was not. It was the end of old-growth because it increased *demand driven forestry* harvesting by over 50%. What's the motivation in this murder? It successfully replaced the railroads who were shrinking due to trucks and automobiles becoming reliable and inexpensive.

Business as usual? I suppose that is the depressing gist of nominal measurement's implementation. Too bad the real environmental cost was never calculated until now. The bill for it today is ginormous and due and the timber barons and politicians who profited, well, their history should be rewritten.

THE AMERICAS WERE LOSING THE LAST OLD-GROWTH FORESTS TO CONSTRAINED AND UNCONSTRAINED DEFORESTATION.

Cities and populations were all growing like mad from 1800 on. After 1920 nominal measurement made all that growth use more forestry, artificially. All of that growth required most of the remaining old-growth forests to be delivered by truck to the sawmill, from everywhere. These included parts of the Amazon, some of which had now entered into harvest rotations while much of it was being converted into farm and ranch land.

All the world's earlier forestry use by 1850 tipped the scale in climate change's favor, but it is nothing like intensity after. It did take 100 years after 1850 before much of the global forest's maturity was depleted of old-growth and replaced with alternate land uses and the immature tree rotations of today. The finale for old-growth occurred around the year 2000. There are exceptions to that year. Within the Amazon Forest it's still happening at breakneck speed.

Ironically, history also records the year 2000 as a new millennium. The year 2000 was supposed to ring in a new millennium maliciously, with a rumored world ending calamity. Something about computer glitches that were going to end civilization, if memory serves. I didn't believe that then but now, well, maybe it did. Perhaps, a slightly slower acting calamity than expected did happen. I mean, loss of old-growth peaked in 2000 as they became rare, protected, and no longer predominate in any forest. Well, that turns out to be a world ending calamity, just not the one predicted 24 years ago.

Not surprisingly, there are people everywhere who insist on harvesting old-growth like it's some kind of personal vendetta or tattoo they need and can be proud of. If they only knew of sequestration sciences' importance perhaps, they'd change their ways. Probably not, and that is another calamity to worry about. The knotted thud of an ogre's sequestration club might do it or you helping to spread sequestrations knowledge. Either works, but one is preferable over the other. You decide which.

LET'S TUNE-UP EMISSIONS REDUCTIONS

The amount of natural and human emissions greatly depends on the data source picked. There are many credible sources documenting emissions. But 435 gigatonnes of CO_2 annually is a reasonable average estimate for the two combined. Because this combined level is so high, I can throw a hand grenade from far away and not a dart and still hit the bullseye accurately. The point to be made is also regardless of precise natural emissions levels or the smaller 35-45 gigatonnes of human emissions. Natural emissions are obviously magnitudes higher. They're also entirely out of any emission reduction attempt's reach. Another of Earth's even larger CO_2 concerns is out of the emission reductions reach. They completely ignore atmospheric CO_2 ppm.

It's again time for a 1,541 gigatonne global warming bomb. You can **skip** the math and continue to the summary if you like, but I suggest not skipping **The Green Conclusion** at the end.

ESTIMATING EMISSIONS' EFFECT ON CLIMATE.

1. Let's say 400 gigatonnes of natural CO_2 emissions annually. I'm going to use the lower estimate for naturally occurring emissions.
2. Let's say 35 gigatonnes of human CO_2 emissions annually. Also, the lower estimate.
3. Let's use 5.1480×10^{18} kg as the mass of the atmosphere. A common estimate.
   a. The atmosphere's composition is 28.97 g/mol, so the atmosphere consists of 5.1480×10^{18} divided by .02897 which is equal to 177.7×10^{18} moles.
   b. A ppm is therefore equal to = 177.7×10^{12} moles.

c. One mole of CO_2 has a mass of 44.01 g, so the mass of one ppm of CO_2 is 177.7 x 44.01 and equal to 7,821 x 10^{12} grams, or FINALLY, **7.821 gigatonnes of CO_2 in one atmospheric ppm.**

d. At the time of this proof (July 2024), the current global atmospheric CO_2 ppm level is 427 ppm times **7.8 gigatonnes CO_2**, which equates to **3,339.6 gigatonnes** of CO_2 that has accumulated in Earth's atmosphere about 0.043% by volume. <u>Way too much</u>, even without CH_4's CO_2 equivalence of 47.75 ppm added in, which sequestration science does affect but not with photosynthesis in this comparison. It is a reason to continue to reduce emissions, BUT? Only because it was an easy assumption to make when not considering sequestrations side of the equation.

e. Earth would like to be around 230 CO_2 ppm give or take 20 ppm. Let's call it 230 ppm which is 1,798 gigatonnes in atmospheric residence. Therefore, today's 427 ppm level is 3,339.6 gigatonnes minus 1,798 gigatonnes = **1,541 gigatonnes of CO_2** that are required to be sequestered from Earth's atmosphere, right here, right now to end global warming. And not considering future years adding approximately 435 gigatonnes annually.

Are you with me? Okay, lets add 435 gigatonnes in!

4. In total, 1,541 gigatonnes of CO_2 in atmospheric residence + 35 gigatonnes human + 400 gigatonnes natural equals **1,976 gigatonnes** of CO_2 that must be sequestered to mitigate or cure climate change today, right now. That is, to end up with 230 ppm CO_2 still in our atmosphere. A cozy blanket. So, that is what it would take to end global warming within one year. Which is impossible. It can't be done within one year because we must start today, and Earth's forestry sequestration is currently broken and that is the one and only thing that balances CO_2. Those 1,541 gigatonnes CO_2 in atmospheric residence are going to take

years of plant growth cycles before maturity can catch-up. And that is the exact reason why we must start CMS mitigation now and not later. We need years for maturity to catch-up.
5. However, removing 435+ gigatonnes and more CO_2 per year with CMS's tools is quickly achievable. Per the earlier CMS mitigation examples adding to current sequestration levels. CMS's first goal is to stop the increasing ppm level. The other 1,541 gigatonnes in atmospheric residence will require 15-20 years. If all goes as planned. But it won't, I already know that and so do you. Anyway.
6. The point is that 1,541 gigatonnes can't be ignored. Unfortunately, emission reductions do ignore it entirely. CMS doesn't ignore it because it can't. Restoring sequestration over time takes large bites out of that 1,541 gigatonne problem every year, like it or not.

Note: 280 ppm could be the more desired atmospheric ppm level. I only use 230 ppm as an example, even though it's a little too close to photosynthesis's lower limit for me, I won't decide the actual number, you will. Anyway, the outcome is the same with any desired ppm level the world chooses. And more importantly, this outcome is reproducible. Basically, emission reductions are lost to sequestration every time, it's not even close.

THE GREEN CONCLUSION.

Emissions-based climate-curing attempts address 1.7 percent of the cause and accumulation of atmospheric CO_2. Less than two percent at fault for influencing global warming's origin or cure. <u>Less than two percent</u>! Okay, admittingly there are two ways to do this calculation.

First way, the numbers applied conservatively: thirty-five gigatonnes of CO_2 divided by 1,976 gigatonnes of total CO_2 that must be dealt with. All of which equals the less than two percent, 1.7 percent, impact that reducing emissions "could" have if that reduction was even possible – but it's not, it's the emission reduction dream and not the reality of sequestrations logic that define *emissions and sequestration dependence.*

Second way, in fairness, it would not be responsible to not mention an alternate way of calculating the above result. This way includes another fairytale. We forget all those CO_2 ppm's in atmosphere. So, humans 35gt divided by human and natural emission, 435gt. However, the result is emissions reductions still only address 8% of global warmings cause and zero percent of a cure. On the other hand, sequestration still takes the responsibility for the remaining 92% of CO_2 that is increasing Earth's atmospheric levels, and 100% of the CO_2 that is in residence conditions now. So basically, Sequestration is the 100% cure for global warming by taking responsibility for 100% of residence conditions levels and all known emission sources. Now and into the future, forever.

In contrast, sequestration has no preference in how CO_2 got there or where it comes from. Which is why I agreed to name the study "Complete Mitigation Science." That said, it is now time to pop the big question.

Having read and seen the considerable difference in human and natural CO_2 emissions' volumes and understanding the enormous amount of CO_2 in our atmosphere and how it gets there, plus <u>seeing</u> the ginormous difference in the two approaches to cure global warming:

WHAT'S MORE IMPORTANT TO END GLOBAL WARMING? REDUCING EMISSIONS OR FIXING OUR FAST CYCLE CO_2 SINKS BY RESTORING FOREST MATURITY?

Lowering emissions is a great and necessary thing to do because of Earth's *carbon hump* and other good reasons. However, CMS is better at curing what we all understand as climate change. As I've tried to explain, it's not climate change it is geoengineered global warming. And, CMS's sequestration knowledge plainly owns global warming's life sucking soul, mathematically.

Here's another logic bomb that kills emission reduction attempts like "Raid" kills bugs.

IF ALL HUMAN EMISSIONS MAGICALLY DISAPPEARED TODAY, GLOBAL WARMING WOULD STILL BE GETTING WORSE BECAUSE OF DECLINING SEQUESTRATION AND NATURAL EMISSION LEVELS. YOU COULD WEAKLY ARGUE IT WOULDN'T BE HAPPENING AS FAST AS IT IS NOW, BUT IT WOULD STILL BE THE ENORMOUS PROBLEM IT IS TODAY WITH ONLY 1.7 TO 8% LESS ACCELERATION.

That one, well, it kind of says it all doesn't it? Even without mentioning that *accumulating decline* adds to natural and human emissions levels every year. Or, that natural part isn't within human control, so it will never "leave" nor can it be reduced without sequestration. Plus, it's a fantasy to believe we'll reduce all human emissions. They have proven to be unavoidable and essential to self-domestication efforts. We are *emissions and sequestration dependent.*

There I went, didn't I. Out on the thin science communication limb to make a point from facts. Now, I'm going to crawl out even further.

If we didn't have fossil fuels, we would have burned up all our trees for fuel. The forests can't support all 8 billion of us using them for fuel to cook let alone our energy needs. That's proven by the number of African communities who have been forced to do just that to survive. Their forests are decimated and still picked at. Nor could we have planted more forests on the agricultural land. Then we wouldn't have enough food or trees. The only answer is conserving and producing cleaner energy as we restore sequestration to offset the energy requirement in human self-domestication. A requirement that leaks CO_2 regardless of a green claimed status. Yes, it is greener, but energy production can never be entirely green.

Nothing is ever going to be without some CO_2 emissions. Fewer people are not the answer either. There were less than 200 million people on this planet when their forestry demand started global warming. Plus, we need more people for the rule of large numbers to work, so we will have emissions from here on, like it or not. *See page 191 for the explanation of the **rule of large numbers**.*

I can't say this enough. Not enough sequestration exists anymore to get CO_2 out of Earth's atmosphere and back where it belongs. But CMS proves Earth used to and can again with forest maturity. I'll add that the excess CO_2 in atmosphere comes from historic human forestry use the last 300 years. We know this because CO_2 sequestration's decline is by far more critical in creating global warming than emissions.

Don't get me wrong, electric or hybrid vehicles, solar panels, and other emissions-based reduction attempts can lower CO_2 emissions and other even worse emissions. Doing so can arguably slow climate change, but as demonstrated not by much. What they do well is make our microenvironments cleaner, so lowering emissions is well worth the sacrifice and effort. Especially, reuse or recycling. That provides an immediate benefit by diluting manufacturing emissions over multiple users. Because the real problem with those other things is you must use emission reduction items for 15-20 years to achieve an actual emissions reduction, just to get over their manufacturing *carbon hump*. Unfortunately, you can never claim it as a global warming cure because none of it lasts that long before it needs repaired or replaced. Both of which add to their *carbon hump*.

For the record, don't blame yourself or me either. I also paid for those things, believing I was doing the right thing to fight climate change. And in three more years, the expensive car I bought 12 years ago will do as advertised, except now it needs batteries so back into the hump it goes. Don't despair fellow do-gooders, all humans have a teacher's note for those previous tests of our good environmental faith: we didn't know of CMS, but now we do! I sure needed that teachers note to pass and move on to CMS. But it's far from over.

Today, entrepreneurs from all over would like to industrialize carbon recycling or its removal with all sorts of profit-driven schemes. Don't get sucked in by the propaganda these schemes deploy. Now, you know better. Trees already do just that with their own solar powered photosynthesis. Without maturity, they can't. You don't need to plug trees in or fill them up at that gas station either. Industrialization of any kind requires energy that releases CO_2 somewhere in its chain of events. In addition, perpetual motion does not exist anywhere but in the plans of some entrepreneurs. Not even in trees. Even renewable energy still has CO_2 leakage points, some more than others. Nothing good for our climate is free nor are current mitigation plans free of the years of required CO_2 repayment timelines. Now is not the time to buy yet another profit-based emission reduction, they can't and don't work without CO_2 leakage. Those leaks only add to ppm and do nothing for the real problem, sequestration.

Time to quote "Peace Sells" by Megadeth, "If there is a new way/I'll be the first in line/but it better work this time!" CMS does work and provides ways for our earlier attempts to become excellent tools now and in the future. They'll work or they can "actually" work as advertised, after CMS sequestration standards are added and forest maturity offsets fix their leaks. And to be clear, they won't work without CMS's involvement.

One reason all this becomes fact is that CMS benefits to our environment continue long before and after the emissions-reduction-based thing is worn out, removed from service, junked, or hopefully recycled. With CMS involvement, most trees maturing won't be touched for hundreds of years, **if ever**. Those trees will be there to absorb more climate sins as humans create better and better ways to reduce emissions, mechanically recycle CO_2 from the atmosphere, get along with the neighbors, or make our lives easier.

Unfortunately, it's time to leave the data pool and sit on a towel because I have some unfortunate news. It would be best if you took it personally. Okay, disclaimer time. I'm not going to hold back, and it is a little 1950ish atomic bomb scary accept this is foretold and not the speculation of World War three in the 1950's. Please remember two things: **First**, all I share are facts. **Second**, I really do hope we can fix this mess in time. As my mom use to say, "we cannot dilly dally." And now I'll reluctantly explain why.

MAKE BAD NEWS WORSE? YES, I'M GOING TO DO JUST THAT

I'm not going to sugarcoat this. I mean, you're still here reading, so maybe, just maybe, I'm being taken seriously. Thank you for that, it has been years of work, so truly from my heart, thank you. From here, please remember that any scary facts I share are not without the hope and the real possibilities that sequestration science provides. But candidly, pulling CMS off is a very real, "maybe." Now, I must explain something that I really don't want to. Actually, there are a few things.

UNFORTUNATELY, CMS'S HELPS TO PROVE THAT HUMANS ARE RUNNING OUT OF TIME FASTER THAN THE UNITED NATIONS CLIMATE PROJECT, OR IPCC PREDICTS. NOW, THE 1950ISH DATUM.

Am I saying I know better than thousands of contributing U.N. science or global climate experts? Not exactly. CMS understandings are recent and coming from a scientist that's gone rogue from emission science, me. They lack sequestration knowledge. Right now, I can only hope they'll see CMS soon enough and beat any perception problems they encounter. I am doing my best to get it out there, but it isn't easy to make a change to what is already accepted emission-based policy and procedures. Chiefly, telling anyone they are wrong is always going to be a problem. My grandfather's curse ensures that.

Thus far informing others has been like flying a paper airplane covered in flames while expecting a good reception at a gas refinery's landing field.

A big problem of communicating CMS is found in today's climate predictions. Like me in the past, the general belief is that global plant sequestration is a fixed value. Science try's to predict the future with that fixed value. That thought process uses the same atmospheric CO_2 outflow each and every year when they form their predictions. As CMS tells us, that outflow is not static, page 46, and has been in decline since humans harvested the first forest. That creates trouble in predictions.

It all ties to forest maturity. The world is still missing sequestration computation in our otherwise complex and precision orientated climate predictions. That makes for an incomplete assessment of global temperatures. Prediction or not. That predicament worsens because past climate predictions provided a false positive to emission reductions. A false positive our emissions tainted feeling believes as truth eternal. We couldn't be more incorrect.

Predictions...

To explain, there are two categories for climate prediction models, hot or cold. On CMS's arrival we get a third category, the Goldilocks model, and our climate is just right! That category is made possible because CMS can predict the end of global warming with forest maturities impact. Which has so far been found by many involved in climate predictions as a deal killer. What I mean is fixing global warming with forest maturity, how dare it even be suggested! So, CMS developed a fourth prediction category to share. The or else category.

Climate model predictions are compared to other climate predictions and then judged to be too hot or too cold in temperatures compared to the average of the other models. There are roughly eighteen or so climate models that I know of. All of which, as a no brainer, predict rising global temperatures. Not surprisingly, none have predicted an actual temperature achieved in fruition and many of them had 30 years of past predictions to do so. That is one of those parts before the fifth part's, page 64, reality kicks in, bad predictions. It is also why the climate deniers have ammunition. So, how correct can current predictions be without sequestration computation incorporated? The answer is unfortunately, not that close. Not even atomic bomb close.

Sequestration is found responsible for the majority of any climate model's CO_2 presence. Not including sequestration therefore sets up all contemporary modeling to be way too cold. If those predictions can be more exact by including sequestration science, should it not be included?

What happens when sequestration is added? I'll just say sequestration is found to be far more significant in global warming predictions than emissions, emissions reductions, ocean temperature, and weather have proven not to be. Still, there is a lot of really great work in all 18 predictions. Some I would even exclaim as incredible in detail. Details I would love to see sequestration science included in. However, they only predict the three bears and not Goldilocks so the story they tell is incomplete. They're all in the or else category.

Currently, they all seem to desire to predict a livable increase in global temperatures. Which to me seems to be some kind of twisted reality; a sought-after goal to make a more popular prediction. For the record, a livable temperature increase? CMS states empathetically it is not possible to predict a livable temperature increase, that does not exist. That is because of the declining forest maturity doing away with sequestration. Which makes temperature increase its acceleration towards Earth's unlivable by animals heat apex. Nobody predicting seems to care about that, yet. Nor can a less than two degree increase in thirty years be predicted by any of them. Not anymore. But they all used to. Which should be warning enough even before we add sequestration's ongoing demise in and make those predictions way hotter. Okay, so what does all this prediction stuff really mean? The popular predictions have not added in sequestration yet, so, here comes the CMS warning.

Predictions that provide any kind of acceptable temperature increase as a result of global warming are impossible. To assume global warming is acceptable in any way, shape, or form is a fool's errand. There are a lot of assumptions and social emission-based agendas within many of the current climate predictions. But, predicting without sequestration and only with emission-based data tells us all 18 models have been wrong and will be wrong again. And not by a little either. Now, add to them sequestrations decline, and we can add the **fifth part's truth and name another category** to rate climate predictions accurately. <u>The extinction category</u>. Because extinction comes long before all 18 of the predictions I've reviewed. So, could I be wrong about all this? I really wish I was, but I'm not. The ppm data confirms this omission in more than one way. They're missing the CO_2 data, sequestration data. And that data affects all of the variables applied in those models, all of them.

I do willfully admit that I could be less than perfect in the precision of CMS's predictions but not the accuracy. I don't have a supercomputer, or any of those other model's algorithms or their guts to work with. So, I don't know the best way to incorporate sequestrations decline into any of them. At least not with the greatest precision possible. I am, however, reasonable, and have Grandfather's curse. So, I can add sequestration's decline into their results easily enough and see their temperature predictions drastically increase. I admit that is not the best way to integrate sequestration. It's also why I don't provide a specific list of predicted temperature increases by year, yet. I'd prefer those atmospheric specialists, programmers, and mathematicians to apply sequestration and improve their model.

Currently, predictions are kind of like comparing apples to oranges. As in emissions based compared to emissions and sequestration based. But as far as CMS predictions being wrong, <u>not a chance</u>. The math isn't that difficult, and sequestration computation can't lie about what ppm level occurred in the past. It' foundation is real and not assumed.

Some of my CMS models are made with sequestration computation that include all known emissions inputs. One of my models can evaluate mitigations authority by adding maturity into a known forest and then globally. Another CMS model uses in and outflows from atmosphere to prove sequestration's deltas over time. It shows the trend of sequestrations decline, its shrinkage occurring over time. What makes this CMS model a standout is it worked to measure and rate sequestration's atmospheric outflows (CO_2 ppm Deltas). It proved outflows weren't static, page 46, and are decreasing. CMS models can do what the emission-based predictions can't. They can prove the beginning, form datums, and predict the end of climate change using sequestration. Remove sequestration from them and they can't do any of that because they become emissions only and a match for today's predictions. Going further, CMS models can add CO_2 emissions into the atmosphere and subtract it with forestry sequestration. Forestry sequestration is another model that uses tree and forestry growth to prove maturity and biomass efficacy. Those prove huge impacts on global warming. And the model does so with physical data collected from decades of independent forestry studies. The point is CMS models are physically kept in reality by empirical data, so they don't assume anything. Well, a little, like emission levels, because we have to. Natural emission levels are to numerous to measure each one accurately.

CMS doesn't have a specific model that can predict future global temperatures. Instead, I rely on relating sequestration to historical temperatures obtained and their ongoing compounding effect found with past CO_2 ppm levels. In essence, I use physical results, and not predictions. These results are used to estimate both a non-CMS future and one with CMS. Although, I can add my results to any climate predictions out there and see sequestrations impact, that is not really my specialty. But I sure see the problem in the prediction afterwards because CMS sequestration predictions come within 8% of actual. Yeah, 8%. So, CMS doesn't really predict, it foretells future ppm level that math can equate into temperature. All because we can empirically measure sequestration.

Sequestration science really is powerful stuff for climate predictions but is not an A listed attendee, yet. Still, I'm refraining from predicting exact temperatures because I lack the highly complex weather models and other expertise needed. But I'm not without sequestration sciences more accurate ppm abilities either. So, CMS can predict ppm and convert it to temperature based on previous obtained ratios.

So, increasing global average temperatures 10° Fahrenheit (+ or - 3°) within 30 years (+ or – 5 years) is a very real and now foretold danger. If we stay on the current emission reduction path while we continue to ignore sequestration's decline, that 7-13° is a range possible over 25-35 years. When or if it does, there will be no returning to any climate normal, ever. The CMS's 1950ish datum setting up contributes to this CMS predicted conundrum.

As a reminder, global temperature is inversely square to CO_2 ppm levels and not directly proportional. As a positive feedback loop warmth creates more warmth which increases temperature much faster than ppm increases. With that, here it is, more bad news made worse.

1950ISH DATUM, THE RUNAWAY HAS LEFT THE BUILDING.

Around the *1950ish CMS datum* humans had firmly established climate changing conditions globally. That resulted in what is interpreted as a runaway greenhouse gas effect. That doesn't mean we can't stop it because we can with sequestration. What it means is the most significant problem humans had already created began accelerating noticeably faster from 1950ish on. Sequestration's decline really took off from the 1850ish launch pad. By 1950ish that acceleration increased and is accelerating even more today. As the name implies, it's a runaway. Some scientists are in complete denial about a runaway effect existing at all. Mostly, I find scientists who'd prefer to semantically argue that if CMS can stop the atmospheric CO_2 buildup, which it can, then I can't call it a runaway. Contrary to that opinion, CMS points out where the runaway effect became noticeable. And not what humans may or may not do with sequestration science in the future. As in, the 1850ish datum point is when humans were first signaled the runaway had started growing powerful legs and was preparing to Usain Bolt away. The 1950ish datum is when it did just that. Not how CMS is going to punish it, hopefully in the future. It is that question of whether we will use it that haunts me into formulating my own definition of the runaway. The not knowing forces me.

    The greenhouse gas is of course CO_2. Around 1950ish CO_2 began running away within ppm measurements. The acceleration's acceleration the runaway demonstrated over time is what makes it a runaway. I don't care too much about sugar coating that fact. I understand why emission science does, it can't do anything about that effect. I suppose sugar coating is to avoid panic or keep us from resigning in a defeat which could result in everyone quitting the climate fight. Inserting CMS into that analysis and the runaway no longer needs sugar coating if we decide to deal with it using sequestration. CMS fixes it, no doubt about it. We can now go ahead and panic and keep everyone in the climate fight with a well-proven way to win. Here's how the runaway became intolerable and does exist.

    From 1850 to 1950, what remained of old-growth in the Americas was being harvested. Even the Amazonian basin, the wildest and most inaccessible place on Earth, became active in answering *demand driven forestry's* call with railroads and then roads pushing deep into South Americas interior and the USA's west coast. Not to mention, Central America's old-growth forests completely disappeared. While Malasia and the Philippines were well on their way to having only immature trees. Around this time is when China's second or third decimation of forestry began, they'd finished by 1970. North America finished by the year 2000, Central America by 1930, South America, well, they're still working at cutting the Amazon forests throat every day. The pacific, they'd finished by the 1960's by rebuilding war torn Asia and Japan. Europe was first to finish in the 1600's. Europeans building empires ultimately are responsible for spreading it globally. As an Example, 5,000 years ago 90% of England was covered in forest. Today, less than 10% remains. All of Europe is like this example.

    The global forestry CO_2 sink or sequestration losses by the 1950's had finished off the Earth's sequestration and emissions balance. The animal to forest symbiotic atmospheric relationship was destroyed. From then on, CO_2 accumulating within Earth's atmosphere became the norm. In fact, by the 1950ish datum roughly 4.7% of all annual CO_2 emissions, natural and human, had nowhere else to go but into atmospheric residence so it began to stack-up even faster. And that level increasing has only gotten faster since. Consequently, CO_2 can stay within the atmosphere for much longer residence times, like forever now. Ever since 1950ish, it's been ramping up in our atmosphere at alarming rates to any intelligent person's observation. And here's the scary thing. That buildup started because it accelerated not by decimals, but by mathematically doubling while the time it takes to double decreases. If that's not a greenhouse gas runaway how else can it be defined? With sequestration's demise the only correlation, I think it's the only way to correctly define it.

TODAY'S PREDICTED 2-DEGREES WARMER IS ACTUALLY TOMORROW'S 7-10-DEGREES WARMER, AND IT WON'T STOP THERE. IT CAN'T. IT'S A RUNAWAY (WITHOUT CMS'S SEQUESTRATION CURBS).

As I explain this part, I'm made uncomfortable. The first global temperature increase sponsored by humans was possibly recorded around the 1850ish datum. It's thought the first one took about 140 years to achieve by increasing forestry demand. The next temperature increase was a double. It only needed one hundred years to appear, and that seems to have happened by the 1950ish datum. The next increase, another doubling, took only 50ish years and happened by the year 2000 (maybe, even in the 1990s). Now 24 years later, there's been another doubling in temperature increase, arguably totaling just below +2° Fahrenheit, 14° C, since 1710 CE. All because CO_2 doesn't have any another place to go. Another double of 3-4° F, 16-19° C, within 10-15 years doesn't take a rocket scientist to predict. The pattern formed and used to predict is easily tied to rising atmospheric CO_2 ppm levels.

But wait! I did say "arguably" because it depends on whose study you read. And why I'm uncomfortable in this topic. The doubling in half the time is consistent throughout all studies, to include mine, so that part is factual, but the average temperature increase is not. They're as inconsistent as a Google search these days, but the majority agree with doubling as they rise over ½ the time of the previous increase. So, because of the extended sample size of science papers on the subject, CMS concurs with the base doubling in half the time. Plus, the average global temperature being higher is agreed to across the board. It's hard to precisely tell how much higher it really is. But it is no doubt higher. I personally suggest it being closer to 3° F then 2° higher along the equator because of the certainty in ppm correlation. In summary, the average global temperature is way too high and going higher! Allowing even higher global temperatures is not acceptable because it won't stop increasing on its own and it will do so at an even faster pace. It is a closed loop that feeds on both CO_2 ppm and warmth.

So, we are in more trouble than any political or current scientific agenda might bravely or honestly predict and then address. Especially in non-sequestration-based climate predictions. It frustrating to know those predictions can't see what's really coming without including sequestration's decline. Why is this happening? They, like you, are just now seeing CMS's data. And belief of new data, yeah that thing, well, that could be a problem even after they're read in. Definitely an or else scenario is now upon the science community and perception is difficult to change. Especially in science. Where these days, your funding tells you what to believe.

So, climate change is a lot worse than previously understood using emissions reduction science. Hope? Yes, there's hope. Do we have the time to fix it? I can only answer yes then tell you I mean **maybe**. Just understanding that sequestration is more critical than emissions is a big step. A step needed to secure hope. Accepting that fact only helps to fix our runaway problem sooner than after it becomes too late.

And remember, hope only grows in well-tilled fields fertilized by mistakes and always in hindsight's rays of clarity. Hope must be planted. It can only then produce the solution to harvest.

ACCUMULATING DECLINES, A SIDE TRIP AS AN EXAMPLE TO A MAYBE. AND YES, I'M STILL MAKING IT WORSE.

I drive down from Oregon to visit family in Northern California often. It's a 6-hour drive. We've been doing that since moving here decades ago. I remember the first trip down there vividly. Snowcapped mountains, Mount Shasta with its glaciers looming, everything covered with immature trees. To my naive eyes it was a peaceful drive, green in the valleys, snow caps looming was not a matter of the season, and the air smelled fresh. Beautiful to see, something I looked forward to. The drive was refreshing, vibrant, and there wasn't much traffic or craziness until we arrived and began experiencing Sacramento's traffic holics. I'm not making fun of you Sacramento; I'm making fun at the situation.

Today, that drive is not enjoyable or pretty. And it's not entirely because CMS ruined my view of the forest. Instead, it's because the land between here and there looks like the burnt end of a wooden matchstick. Large forest fires made that happen over the last several years. It's a noticeable part of the *accumulating declines* the global warming trophy is made from.

That drive smells of charcoal and is as hot as a desert tortoise's back at noon during a Texas summer. There are no snowcapped mountains unless it's late winter or early spring, and Mount Shasta's glaciers are hardly noticeable because they've shrank into oblivion or have vanished. It only took sixteen years for all that land to morph into the form it's taken today. Much of this happened in the last ten years. All the residual effects of tree immaturity make global warming a combination of terrible things that make each negative effect worse. And those effects are not just accumulating, their occurrences are accelerating in step with atmospheric CO_2

levels. The downward spiral is increasing speed as it takes us closer to the abyss.

Climate change's temperature increases influence forest fires. The result makes fires more and more likely due to drought. Drought is of course brought about by increasing global temperatures. And gets its own section of explanation in this book later. For now, global drought increases fire seasons as heat waves suck the moisture out of forests. Today's forest fires burn hotter and cause soil to become sterilized. As in nothing can grow there until the soil is restored. Now let's talk about bugs. Specifically, beetles.

Because the average temperature has risen enough some forest killing beetles have been able to expand their range into much higher elevations. Replanted nonnative trees also assist beetles to spread. Beetles and other wood boring insects are killing entire forests; and one answer to why I answer maybe it is too late. They are spread across the entire globe and boring into trees and laying eggs where they shouldn't. The eggs, nests, and wood eating offspring destroy the tree's vascular system, its growth ring, and thus the tree's ability to water and grow. The result is the beetle infestation eventually kills the entire forest by turning trees into erect piles of kindling. Fire that sterilizes the soil is almost certain to come next as the canopies the trees used to block the sun are gone. That allows understory plants that are fuels to grow in abundance. It is only a matter of time before the fire destroys it all.

The beetles don't mind killing the tree because the dead tree becomes even better habitat for their offspring to spread to alive trees the next season and start the process again. These beetles know all the globe's languages, so they don't care what country they create havoc in. All because the average temperature has increased enough to increase their range. And man's preference for one tree species over another that isn't native to the region was replanted. One of many *accumulating declines*. One that in turn further helps fire consume vast amounts of forest, among other reasons just as nefarious. And there are so many *accumulating declines* I lose count.

More examples, immature trees overcrowding forests because entire ranges were clear cut and so they are in regeneration at the same time, and all the time. That practice makes a book of matches and hands them to children. Or, more commonly, natural lightning will at some point ignite those forests like a firework fountain on the 4th of July. I've seen seventy-four acres of 12-year-old replants disappear in a few minutes, personally. Like a small nuclear weapon was dropped, as I recollect. The mushroom cloud formed from the carnage was emense.

Add in the rise of forest fires threatens the last old and mature growth stands globally. Old and mature growth is very resistant to fires; unfortunately, the highly immature tree stands surrounding them burn like lighter fluid. Because of immature trees fires can and do penetrate the tiny acreages of old, mature growth stands, and destroy them as well. To clarify, it happens because trees surrounding them are immature, hold little moisture, and burn unnaturally hot. Not because Old and mature growth burn as well, they are wood and can but that's the exception in healthy forests not the rule. Therefore, naturally old-growth doesn't burn-up anywhere near as much as immature tree plots. Not without the kindling immaturity provides.

Again, all the *accumulating declines* make me worry about the global future. They are accelerating global forestry towards an extinction that will arrive before our own. A world without forestry is a world without hope.

CMS is forced to work with what we have left. We must engineer forestry into what we need given the tools available. Except this time our geo-engineering won't ignore CO_2 *sequestration value* or biomass efficiency. CMS's mitigation efforts and I plan to go down fighting, if Mother Nature and humans do not cooperate, but she'll have no choice if we do.

We will resurrect Titans and take an accountable control of our surroundings.

Okay, Mother Nature is not known for her cooperation. In fact, I'm pretty sure she's okay with us getting hit over the head by the global warming trophy overflowing with *accumulating declines*. At this point, I'd even say she'd enjoy providing us a beating. She has for sure put a time limit on humans fixing our geo-engineering and reversing our bad forestry management. When ignored, she does become intolerable.

Physically fixing my trips drive will take time. The time needed is 100% governed by nature. We must play by natures rules because those rules are no longer negotiable. Fortunately, CMS's reverses all *accumulating declines* in forestry with nature's very own specification on tree maturity. Nature is here to help even while Mother Nature warns us it's her way, or else.

I introduce numerous *accumulating declines* throughout this book. As you read the list you'll probably realize how one decline can create a chain reaction that creates and accelerates others, their all tied tightly to number 1. There are more than listed.

[1] Tree maturity decreasing
  a. *Impeded CO2 fast cycle sinks*
  b. *Tree and land degradation*
  c. *Constrained and unconstrained deforestation*
  d. Decreased Atmospheric CO_2 regulation and balance
[2] Human geoengineering
  a. *Demand driven forestry*
  b. Old Growth Tree loss
  c. Harvesting of healthy immature trees
  d. Land use change
  e. *Carbon Humps*
[3] Rising atmospheric CO_2 ppm
  a. Decreasing atmospheric CO_2 outflow
  b. Global warming
  c. Extended CO_2 Residence conditions
  d. Human and photosynthesis PPM limits
[4] Global average temperature increase
  a. Ecosystem modification
  b. Ocean warming
  c. Artic melting
  d. Storm intensity
  e. Weather patterns
  f. Livable conditions decline
[5] Drought and weather-related declines
  a. Slower plant growth rates
  b. Storm damage to ecosystems
  c. Decreased crop production
[6] Excessive CO_2 fertilization
  a. Instability in plant cellular growth
  b. Decreased plant life cycle.
  c. In adequate water intake.
  d. Acidosis, plant death
[7] Ecosystem modification,
  a. Beetles and bug habitat expansion
  b. Decreasing forest tree species habitat
  c. Decreased animal habitat
[8] Forest Fires
  a. Increasing number of
  b. Increasing intensity of
[9] Our view of the forest

TIME TO MAKE THE WORLD BETTER THAN WE INHERITED IT

Let's eliminate **all CO_2 emissions** using mature trees. Since we can. Knowing what we know now, it really isn't that difficult, or is it? That depends on whether you're still enslaved to the dark side's emissions-based empire or joined the rebellion and using sequestration's force.

My apologies for the math. A ***SKIP*** to item 6 is possible.

[1] Because humans need to remove gigatonnes, **let's** see how many Lbs. are in one gigatonne. First, 2,204.6 Lbs. (0.45 Kg) are in a metric tonne (1,000 Kg). "Giga" means one billion (1,000,000,000), so a gigatonne, gt, is one billion tonnes. So, 2,204.6 Lbs. per tonne times one billion equals 2.2046 E^{12} Lbs. in one gigatonne is the same as 2.2046 x 10^{12} and is rounded to 2,204,600,000,000 as two-trillion two-hundred and four billion six hundred million Lbs. in a single gigatonne of CO_2. **That's a lot considering we are referencing one CO_2 molecule that weighs 44 amu into a billion metric tonnes!**

[2] So, humans produce something like 35 gigatonnes CO_2 yearly times 2.2046 E^{12} Lbs. of CO_2 = 7.7161 E^{13} Lbs. of CO_2 to be sequestered.

[3] Here's what today's *constrained deforestation* currently does with its estimated paltry 1,100 Lbs. of CO_2 per acre annually sequestered with immature trees, on a good day. Keep in mind the *carbon hump* and immaturity create this 1,100 Lbs. estimate. Its likely much less per acre. It certainly isn't going to average higher than 1,100 Lbs. after the *carbon hump* and immaturity have their way.

[4] As a reminder, 8.5 to 9.5 billion acres of forest (out of 10-10.3 billion global forest acres) are unprotected and within

constrained deforestations definition of practices from *demand driven forestry*. We are not counting protected acres, yet.

[5] So, 8.5 billion acres x 1,100 Lbs. of CO_2 sequestered per acre = 9.35×10^{12} globally sequestered Lbs. of CO_2 in one year, said as 9,350,000,000,000 (nine trillion three hundred and fifty billion Lbs. of CO_2). And now Lbs. to gigatonnes. 9.35×10^{12} divided by 2.2046×10^{12} equals an estimated an paltry **4.24 gigatonnes** of CO_2 being sequestered under todays *constrained deforestation* using the Earth's unprotected forestry. That's right now and not good at all. Again, along with immaturity, the *carbon hump* is more than partially responsible for that low number. What I mean is that **4.24 gigatonnes** are mostly due to trees emerging from the *carbon hump* just before they are harvested. The remaining trees don't sequester enough CO_2 to be useful in *sequestration value* or in the longer version of this calculation. They are essentially nullified after considering their immaturity and position in the *carbon hump*. I need to say this is an estimate that lacks precision but is correct in assessment. They don't do enough; we don't let them.

[6] In comparison, the last remaining mature forests and old-growth globally are said to occupy around 1.8-1.3 billion protected acres, roughly 15% of the total forest area of 10.3 billion acres. That again depends on the source; I'm using United Nations data here, and don't believe it as I explained earlier, maybe 750 million acres truly exist with old growth trees but even that is questionable in data sources. For the record, old-growth holding acres are off the sequestration charts, as you'll see shortly. I mean one tree can sequester the same amount as one acre of immature trees and some species even 4 times as much. Because they're so mature, 300-plus or even thousands of years old. They can sequester tens of thousands of pounds per acre. Again, being mature does not automatically mean

they can. Species and location also play a large role. However, marketable tree species, well, they pretty much all can with maturity.

   a. I estimated they could be sequestering around 2.032 x E^{14} Lbs. of CO_2 annually, as in 203,200,000,000,000 trillion Lbs., or simplified as **92-96 gigatonnes**, of CO_2 sequestered yearly. Which is obviously not enough given the constant rise in atmospheric CO_2 ppm levels. Now, I've also applied a *carbon hump* to them as well because in a perfect world scenario they should be sequestering **150 gigatonnes** yearly and even up to **220gt** without it. But thats not what's happening today.

      i. Applying atmospheric ppm to the globe's sequestration measurement and they, by themselves and with all other sinks combined, fall short of balancing the atmosphere, by between 20-30 gigatonnes every year. Which says either the *carbon humps* impact is not limited to *constrained deforestations* excessive emissions, or the old-growth present is less than publicly stated. Personally, I believe it's some combination of both and more. The hump might last for decades longer than I've stated and maybe even centuries because old-growth is absent. OR excessive CO_2 fertilization has damaged both old-growth and newer sinks, there weren't enough old-growth sinks to curb that *accumulating decline's* impact from becoming worse. And, as I've explained, data available on their existence is at face value inconclusive and requires interpreting to sequestration's higher data standards.

[7] The two examples combined, let's say that 96.24 gigatonnes of CO_2 per year are sequestered by forestry, cleanly and after the forestry harvesting hump is deducted. Not even close to their potential. It appears as other sequestration sources like the oceans and other plant life are doing most of the estimated 435gt of Earth's annual

sequestration these days. Of course, we know even all of that helping still falls short by 20-30gt every year. Forestry has almost bowed out of the task at hand and is creating global warming's trophy as it leaves the fight.

PREDICTIONS WITH CO_2 PPM.

[1] Moving on, one atmospheric CO_2 ppm is estimated at 7.82 gigatonnes CO_2.

[2] The ppm level from May 2022 to May 2023 increased by 2.8 ppm as averaged. That is equivalent to 21.896 gigatonnes added to Earth's atmospheric residence. CMS's first prediction using the trend proven within CO_2 deltas missed that increase by less than 1.7 gigatonnes in 2023. It was within 8% and worth mentioning.

    a. The next prediction was for May 2023- May 2024, a 2.9 ppm increase was measured at 22.678 gigatonnes. Again, CMS is within 8% of its prediction.

    b. The 2024-2025 CMS prediction is 3.26 ppm, 25.024 gigatonnes.

    c. 2025-2026 prediction is 3.71 ppm.

[3] I think you can see the pattern, the decline of sequestration, in the trend I mentioned earlier. That is 1974-2021 CO_2 ppm Delta, as $Y=9.56$. Which is proportional to predicting annual average atmospheric ppm increases with some added math applying to its increasing base.

    a. I can propose better precision, as in being within 4% of predictions in the future because there are better ways of building this model with more variables. Within 8% for now is better than I hoped for.

[4] What is significant within ppm examples? First, they provide a second proof that sequestration is in decline, because emissions didn't increase enough to cause the year-to-year ppm increase. Something else is happening, sequestrations decline. It also tells us that any improvement

in sequestration is a winner-winner chicken-dinner for curing global warming. Which also relates why CMS offers the only global warming solution by explaining the increase. Forestry must be allowed to do its part, or else. As to the predictions, within 8% is pretty good in any prediction and twice in a row tells us we know what we're doing.

CONSERVATIVELY, LETS FIX WHAT WE BROKE:

So, I'll use a tree at age 59, the ability of that tree is approximately 134.1 Lbs. of carbon sequestered into its mass from sequestering 492 Lbs. of atmospheric CO_2 in one year (the 3.67:1 amu ratio).

[1] Now, I need 109 trees per acre because they're older, bigger, and need more space than the 30-year-old version did at 190 immature trees per acre which started out at 240 replanted trees per acre. I think you can see natural attrition occurring over 59 years with that statement. Anyways, 109 trees per acre x 492 Lbs. CO_2 per tree = 53,628 Lbs. CO_2 per acre sequestering ability. Wow! A lot better than the current global average of 1,100 Lbs. per acre all by itself. Maturities sequestration power difference already!

[2] Now, 8.5 billion acres x 53,628 Lbs. of CO_2 per acre = 206-213 gigatonnes of CO_2 sequestered from the atmosphere by increasing tree maturity up to 59 years in the unprotected forest. BUT WAIT FOR IT... NOW, add those 206 gigatonnes to the 96 gigatonnes of existing mature and protected forestry, and what do you know! 298 gigatonnes of CO_2 are sequestered into forestry. This is in addition to Earth's other CO_2 sinks, like oceans and other plants and even Earth's slower CO_2 cycle of geological sinks. Anyway, you should get a warm fuzzy feeling. I define that feeling as more hope in being able to fix global warming quickly!

[3] I have even more hope to apply! Remember the carbon hump? Every year replants and natural regeneration in forestry do manage to grow past the carbon hump. And so, they at once start to suck CO_2 out of the atmosphere instead of just their surroundings. In effect, sequestration restoration becomes even better than exponential and pardon the pun, "log"-arrhythmic.

[4] And one more thing, replanting forest land in afforestation efforts (restoring land that was once forest) has little to no carbon hump if it was deforested over 20-30 years prior. Yea! I can go tree planting on used to be forest land and will! But wait, you still need patience to achieve sequestration value. Remember, natural attrition found within afforestation makes its own carbon hump. Darn it! There is no trick to getting out of waiting decades for afforestation efforts to gain sequestration value! But, make no mistake about this, its worth the wait!

NOW LET'S TAKE SEQUESTRATION TO A WHOLE NEW LEVEL.

[1] Let's apply an average 6% annual tree growth rate using the 8.5 billion acres (Typical tree growth rates are 3 to 15% per year). I'm dropping to 79 trees per acre to make some room for more mature trees, which optimizes sequestration by growing more limbs. We'll also add 96 gigatonnes of the existing mature forest even though it includes the current immature forest's gigatonnes. Keep in mind, as we resolve *carbon humps* and *accumulating decline* with maturity that 96 gigatonnes grows logarithmically beneficial but is not included in these examples. I want this point made without that beautiful distraction of their increasing sequestration value. The point is the improvement over Earth's current dismal

situation is by using only one part of sequestration's improvement.

[2] At age 66, 225 gigatonnes of CO_2 are sequestered per year by maturity improvement alone. Natural attrition is answering modern and even increased forestry demand. And we've started extracting excessive CO_2 from the atmosphere.

[3] Age 71, 301 gigatonnes for a total of 393 gigatonnes CO_2 sequestered per year.

[4] Age 78, 453 gigatonnes for a total of 545 gigatonnes CO_2 sequestered per year. Note: here is where we are sequestering all emission sources and rapidly extracting from atmospheric residence with forestry alone!

[5] Age 84, 643 gigatonnes for a total of 735 gigatonnes CO_2 sequestered per year. We can now prove another CMS datum. A datum the tells us when global warming ended. The 1,541 gigatonnes in atmospheric residence is almost gone. Earth will be around 280 ppm and global warming inspired droughts outdated. Ah, if it were only as easy as the numbers tell us. It isn't, but at least we have proved forest maturities impact, again.

The best part, sequestration and biomass efficacy continue to improve with age beyond this example. And it's why I don't care exactly what natural or human emission gigatonnes really are (400-750gt). Because sequestration science doesn't necessarily care where emissions come from, it only cares about where to put them, permanently. And sure, I'm providing a perfect world scenario to prove sequestrations point. So, allow me to offer the following to any still doubting: If only another 3 billion acres were allowed to mature 75 years beyond average harvest age, we will have not only balanced sequestration and emissions we'll have removed excessive CO_2 from atmospheric conditions along the way. We'll oversee

Earth's thermostat from then on. Give me four billion acres at 55 years old and we can do it in around half the time.

Plus, the biomass efficiency of those mature acres would be triple that of all the remaining immature acres. And the added acres could supply forest demand all by itself with natural attrition providing the thinning harvest all the while increasing maturity, and forest health, soil health, water quality, and habitats, and our view, and weather, and an end to global warming. Hmm? Why not? Caveats, the right regions, species, and trees over the *carbon hump* to start with please.

And that brings me to a point of frustration. Not because I want to be the spoiler, it's because you need to know.

Even if CMS has been mitigating CO_2 for a decade or even more, humans will still be in greater jeopardy from a much more natural and recurring problem interacting with our current global warming trophy. Yes, scary stuff is getting in the way and why I say "maybe" while sequestration screams, "or else." Unfortunately, hope can quickly turn into desperation, just as it has and could again.

IT HAPPENED BADLY BEFORE; STOP MAKING IT WORSE

Kaboom! The Impact of Volcanic Eruptions on Climate Change

In 536 CE, 1,487 years before this book and 147 CO_2 ppm's ago, the Krakatoa volcano erupted, changing humanity forever. This eruption created climate chaos, setting human self-domestication back centuries. Although some theories suggest multiple eruptions, scientists studying tree rings and geology have identified Krakatoa as the main, if not sole, culprit.

The exact volcanic details remain uncertain, but the aftermath is well-documented globally. Historical records mention days turning into nights, ice-cold temperatures, and crop failures to cold and zero sunshine. And then the worst part happened. Unprecedented decades of droughts with crops dying of thirst. The 20-years of combined global impact may have reset societal behaviors, many historians believing it did just that.

Why is a volcanic eruption significant to CMS? Because the 536 CE event would have been far worse with today's 147 ppm higher CO_2 levels. I view volcanoes as common and their occasional eruptions as expected. My own experience witnessing Mount St. Helen's eruption's ash fall in Montana taught me the importance of respecting volcanoes.

Today, Earth's volcanoes are more concerning than asteroids. We have technology to detect and handle asteroids, but nothing for volcanoes. Preparation is essential, yet global warming complicates this. Even the best "the world's going to end" preppers aren't ready for a large volcanic impact.

A big eruption or two mediocre ones today could leave an indelible mark in history. A mark we really don't won't. All because human recovery with current climate measurements, well, we've made that questionable. Basically, if 536 CE's volcanic event repeated now, the results would be far more catastrophic then humans experienced then.

THE DARK AGES ARRIVE.

After the 536 CE Krakatoa eruption, a colossal tsunami was sure to have ended the lives of many locals, but that death toll was nothing compared to the carnage created globally. The volcanic dust clouds and CO_2 emissions took far more lives than coastally confined waves. The environmental effects were far more costly. They were because they were longer lasting and discreet at killing humans slowly.

For the record, today, all volcanic activity combined doesn't outproduce humans in CO_2 emissions annually. Which is a good thing. But that statement is typical in most years during the last 130 years. A statement made in that geological timeline equates to an unborn babies birthday. It's tiny in comparison to the years required for volcanos to grow and then erupt. However, 131 to 220 years ago, that statement was not true at all! Anyway, massive eruptions like the one in 536 CE emit enough CO_2, ash, and sulfuric discharges that do change global climates, temporarily as the norm. That is until now. The scary problem with Earth's volcanic history is their eruptions emit vast quantities of everything bad for the atmosphere and do it all at once. You can measure their enormous emissions in hours, days, and weeks.

CO_2 levels increasing in the atmosphere is not solely due to the eruption's emissions. Sure, the eruption emissions can, of course, be gigantic. The other reasons that hugely increases CO_2 in Earth's atmosphere, you could say, well, they are just not readily available without study.

Thanks to studies on tree rings and historical accounting from across the globe, we know the world went through 1.5-3 years of a mini-ice age. The duration depends on where on the globe you live; the further north or south from the equator, the shorter the mini-ice age. That is counter intuitive, isn't it?

You'd think the closer to the polar regions the longer the ice age. It just goes to show how large eruptions affect the atmosphere, doesn't it? Anyway, we know this because trees living at that time grew little, if at all. Photosynthesis without sunlight is impossible, so it stopped or suspended during the three years of blackened days. There was little to no sequestration of any CO_2. Many trees died thanks to the suspended dust. Crops didn't fail, they never even sprouted. Thinking about that, people didn't have growth lighting back then so little to nothing grew on the entire planet for up to 3 years. "Everything only withered and then died of the ice or the dark." That's a quote from a French Monk during this global crisis.

Massive amounts of CO_2 were also released during the eruption(s). But that single CO_2 burp was compounded with another massive CO_2 release, and here's the not so easy to think about: the mass dying of plants and animals (including humans) added to atmospheric CO_2. And without sequestration working the excessive natural CO_2 emissions all piled up and increased atmospheric residence conditions. Does that scenario sound comparable to today, without a volcano helping? Even exactly like today's ultra-high 427 atmospheric CO_2 ppm level? And yes, this can be seen in ice core samples, so it is a CMS sequestration proof made by what happens next.

By 538-539 CE, our predecessors looked at the sun for the first time in years. They knew they had survived by being murderous thug's that worked for a ruthless pope, warlord, dictator, or king who was able to hoard food like precious metals are today. These tyrants only fed the loyalists who served at their will. Sorry, I just had to say that about our predecessors while remembering we are no different but have better tools. I'm sure some of them were nice and some vile. I also know some passed on legacies that still impose the cult like infections they created even today. I suppose human loyalty applied in the form of henchman is a well proven attribute and a generational affliction. An ongoing affliction that education prevents. Enough said, that rant is over because nothing can be done so long as people are willing to be henchmen or blinded by propaganda.

By the end of the mini-ice age, nobody had been living well, if at all. Just enough of everything survived the viciously cold and blackened days to make a decent starting over. "The Dark Ages" title really defines those times well. A lot of human civilization collapsed, vanished, or hibernated in secrecy for years and even decades. That's like saying all government disappeared because the people in government were too busy trying to find food for themselves. Total lawlessness as anarchy ruled the need for food. Very little of anything lived through it, and that's because when the mini-ice age ended, the survivors' troubles only warmed up with the appearance of the sun. But humanity did survive. A recovery began; but it was a short celebration.

When the mini-ice age stopped hiding the sun, it was followed by a short timeline of massive floods, then 10 years of catastrophic global drought in the Northern Hemisphere and approximately 20 years in the equatorial regions. Those droughts killed more people than those dark days of cold. Now imagine this: just as the sun comes out and you think it's going to be all right; your crops are first flooded out and then you don't receive enough rain to grow crops for ten or more years. Add that they had no idea why all that was happening and possibly the reason some religions really took off in their numbers of followers!

You might be asking what all the Dark Age stuff has to do with CMS. Well, eruptions have become an even more significant problem when they happen again. Here's the understandings science creates by categorizing all the eruptions effects. Let's start with the flooding. That was caused by the sun appearing and thawing everything out and all at once. Which put a tremendous amount of water vapor into the atmosphere that resulted in biblical rain events and durations. The Earth was still cool so water vapor in the atmosphere couldn't stay in residence conditions very long. It's a condensation thing. And for the record, those rain events were massive, think thousand-year and not hundred-year floods. The next thing Humanity dealt with is global drought. The excessive CO_2 buildup from the eruption, the die offs, and natural and human emissions all contributed to form the perfect dried out storm of dust. With almost zero global sequestration for three years the CO_2 levels globally increased. The atmospheric CO_2 ppm level then was not anywhere near as high as today's. Anyway, Earth did not and could not absorb any of it until photosynthesis recovered. So, up to three years of natural emissions went straight into residence conditions along with the eruption's ginormous contribution. Again, do these results sound familiar? Like today's 142 ppm higher level with our own decades of droughts? So why aren't we starving like they did?

STARVING IS RELATIVE TO TECHNOLOGY

To about 30° longitude north and south of the equator we most assuredly are starving. In those latitudes dryland farming is a rarity and associated with microclimates. It didn't used to be. Today, desertification brought about by drought and *accumulating decline* insure global warming's starvation impact. However, we've invented a lot of tools that hide drought's impact well. Tools not available in 539 CE like, global distribution, electrically pumped sub surface (deep) wells, dams and reservoirs, desalination, and huge irrigation systems that cover multiple states and even entire countries. We've learned how to move and store water thanks to ongoing drought. Unfortunately, even those tools are losing their sharpness as population increases, global warming worsens, and ground water becomes less available for crops.

After sequestration recovered around 540 CE, a growth cycle finally occurred. From that point it still took 10-20 years to sequester the excess CO_2 released. That first plant growth cycled allowed suspended water vapor in the atmosphere to cool and fall as normal precipitation that ended all the droughts. That is how floods begin and droughts end, cooling. Global warming's heat suspends more water vapor which increases the temperature during the daytime and that suspends even more water vapor (as humidity). Essentially, Earth requires the coolness that outer space brings each night in order to regulate our climate's heat buildup during the day. Just not too much coolness or warmness. We need Goldilocks blanket, just right. One leads to global warming, the other to ice age and both equal our extinction. Warming more so than cooling does.

The reason it took longer to end drought along the equator is the eruption was nearer the equator than Earth's poles. So, way more CO_2 existed there than further north or south. The Coriolis effect of Earth spinning, and the sure geographic size (as a much bigger diameter) makes the equator area a magnet for CO_2 buildup and more prone to global warming's effect or events like the one in 536 CE. So how does this all relate to today's climate problem?

Have a look at Figure S1 in the back of the book again, page 229. Looking at the ppm line I bet you can see 536 CE volcano ppm increase. Hint, it's the highest point just after year 0. And only around 270ish ppm. Again, today, Earth is at 427 ppm. So, the 536-drought ended, and our global drought perpetuates.

DROUGHTS...

This CMS assessment lends itself to what the major problem is today, drought. A symptom of global warmings temperature increases and the *accumulating decline*. The first thing we learned about drought comes from the hindsight provided by our past volcanic experiences. Prior to the 1800's, drought was usually the result of volcanic activities releasing significant quantities of CO_2. Temporarily overwhelming sequestration like it is today. The event in 536 CE and the others in this book are recorded as the causes of drought. However, back then droughts were very short lived compared to today's. Only bumps in the road that ended fairly quickly. You see, before and for a thousand year's after 536 CE's event the world didn't see the prolonged droughts we have today. With good sequestration, some ended after a couple of years, other more serious ones in 10-20 years. The recovery was highly dependent on the volcanic event's discharges. The distance from the volcano and the distance from the equator also prolonged the inevitable drought that followed. The pre 1800's droughts also lacked the severity of today's prolonged droughts. Therefore, sequestrations decline increasing CO_2 ppm is no doubt the cause of today's permanent droughts.

Today's horrific droughts began developing around the 1800ish and they will last forever and get worse if unchecked. Some droughts have plagued the world for well over a century. Some along the equator even longer. All this seems to correspond to CO_2 levels exceeding 270-280 ppm. To explain, the 536 CE event happened when the global fast-cycle sinks were reasonably healthy, and the atmosphere had a low CO_2 ppm level in that 270-280 ppm. Sequestration in Europe, Asia, and Africa was damaged but nothing near the damage of 1800 on.

So, what would happen if the 536 CE event happened this week? I don't think a degree in rocket science is required to put that together. I do think setting the stage with questions might help understanding how we will be held accountable at the next big volcanic eruption.

[1] Do today's immature and impeded by millions of percents fast-cycle sinks provide any comfort knowing the 536 CE sinks were millions of times less impeded than now?

[2] How does the current 427 ppm CO_2 level in Earth's atmosphere make you feel? Did you compare it to 536 CE's 270-280ish-ppm level, a level that took healthy sequestration 10-20 years to recover and end the 536 CE event's worst impact, drought?

[3] I'm concerned enough to write about this, are you okay with knowing it? I'm really not trying to scare you. I just want to inform you.

[4] Now the real question. Could we recover if the 536 CE event occurred this week? Consider these factors:

   a. The resulting droughts could finish farming within 45°-60° latitude from the equator. The world's food supply could be gone, permanently. Even now, from the equator to around 30° latitude has gone into desertification mode if not already there, so another 536 CE event could make even higher latitudes impossible to recover a norm even with the irrigation tools we've invented. Look, desalination, and ground water pumping require energy, which means CO_2 emissions, which would make it worse, not better (without healthy forests). Even making and supporting solar panels releases copious amounts of hidden CO_2. Plus, you need sunlight for solar to work.

   b. Wars for food and fresh water will ensue. Food would not be distributed by economies as it is today. If a country doesn't grow enough, it won't be able to buy it anywhere. And forget donations. Like oil is becoming today, we'll have to go and get food by force, if we can find it. By the way, this is already happening where people can no longer grow enough food to support themselves due to decades of drought. Without sequestration fixed, food could replace money as the global currency. And forget globalization, that will collapse to dictator's aligning food resources to stay in power. Like it has already started to do, oligarchs. I'm afraid the brutish regimes are rising, it's do that or starve, just like the dark ages reborn, that is today's globalization of trade that is declining as we speak. Just ask China, Canada, and Mexico if the U.S. is buying as much as it did 10 years ago. Nope, we aren't. The stuff costs more and we're getting less of it.

c. During winter, survivors could take tropical vacations in the Bering Strait of Alaska. Except…Rising sea levels will displace 85% of the entire world's population and they'll move there to escape the heat. Expect crowds.

d. Forestry might only exist in memory as the drought and other *accumulating declines* end it. Our final chance to fix global warming is gone with it. Yes, a large enough eruption will accelerate all of the *accumulating declines* exponentially. But only after the freeze kills the immature trees off. The mature ones will survive.

e. After the thaw, a rated high enough Sunblock isn't possible. You'll know exactly what Bram Stoker's vampires feel when the sun directly hits you. That's due to the magnification caused by excessive atmospheric water vapor in residence conditions intensifying every day and now happening forever as CO_2 ppm runs away even faster. No cooling is no rain. And remember, medically 1000 ppm is certain death for animals and humans anyway.

I am sorry because it's not over yet. I really wish it was, but I can bring this much closer to home, twice more.

536 CE's culprit Krakatoa had another eruption in 1863. Thankfully, a minor eruption. Most research says it added to

the following decade of CO_2 ppm increasing which added to our existing global drought. How much CO_2? Unfortunately, that's highly debatable and one of the reasons I chose the earlier date of 1850ish as CMS's *climate changing datum*. Yes, CO_2 spiked up, but not by a particularly concerning number because sequestration damage wasn't as bad as today's. Plus, by all accounts, 1863's eruption was a much smaller eruption than 536 CE's. In addition, another eruption from a different volcano may have even added to Krakatoa's emissions the year before. Still, the effects were nothing like 536's. So, what, right?

To the point, volcanic activity from the late 1700s to the 1880s was tremendous, much more than we've seen in the last 130 years and we know it is far from over, its only resting. So, here's my take on eruptions and my concern for the record.

Geologists rate eruptions like earthquakes using the Richter scale but add in many other factors. Too many factors too really get into here. To me the VEI scale, Volcanic Explosivity Index, is like scoring gymnasts at the Olympics. There's a lot of opinion involved. That and the 536 CE eruption is an ongoing project for geologists and historians. It's a more recent discovery in the cause of the mini-ice age that came via combining historical accounting with a great tree ring study. Its effect, well it's not as well documented by geologists as it is by historians, but it is recorded in ice core samples which tie it all together. Statistically, it's estimated to have been a high VEI 7. Because of the well-recorded history of the carnage that followed it gets a higher ranking than one that doesn't cause interpretable degrees of carnage. Consequently, it doesn't rank high enough in VEI scale to even make the top ten eruptions on Earth. Yeah, all that global carnage and not even in the top ten. See what I mean about the VEI scale?

Now the later 1863 CE Krakatoa eruption is documented by geologists and considered a VEI 5 or low 6 by some, so it was more like a countries coal burning electrical facilities working normally for a few years, except its CO_2 emissions released all at once and did add to global drought. But not a mini-ice age or twenty years of drought. Because this occurred after the 1850ish datum, technically, CMS considers it to have helped tip the scale into today's global warming. However, the 536 CE eruption was by far more destructive and a total of 2 magnitudes of 10 higher in rating. So, one magnitude is 10 times more powerful. Two makes it a toaster to a commercial kitchen in comparison.

Our global CO_2 fast-cycle sinks are also two magnitudes from before the 1863 CE eruption(s). They were also almost stellar prior to the 536 CE eruption, but not perfect by any means. Humans began geo-engineering forests eons before, but those results really only began showing noteworthy atmospheric ppm accelerations after the 1850ish datum. The point is, in 536 CE there weren't anywhere near enough humans globally to challenge forestry with globally applied *demand driven forestry*, yet. Europe, Asia and specifically China were exceptions, but they even had a lot of old-growth in remote corners everywhere because of a lack of transportation technology. Yes, we had been practicing *constrained deforestation* for thousands of years before 536 CE. And we were logging all over the globe by the 1863 eruption. Still the Americas, Africa, and South Pacific were "mostly" untouched. And yes, some Indigenous peoples were tough on their forestry with fuel use, fires to clear land, and even temple building, but not to the extent of Europeans and Asians!

Here are points to think on:

[1] The atmospheric CO_2 level was an estimated 280ish ppm level around 1863. In 536 CE, the level was approximately 276 ppm. And lots of mature and old-growth trees were still all over Earth. Especially in the Americas.

[2] Another 536 CE Krakatoa-like eruption today could change the current 427 ppm upward in a massive way— 10's of ppm in a concise period, like a week. Closer to the volcano, it could be much more. And very limited old and mature growth globally supporting our recovery. Our trees, to include old-growth, are in Earth's *carbon hump* already, so they won't help in our recovery. But they did in 536 and 1863.

[3] Humans are full of carbon. It's 18% of your mass, the second most abundant element in human mass. One 150 lb. human is 27 Lbs. carbon. If cremated, that 27 Lbs. turns into 99 Lbs. of CO_2. Burials allow the carbon to enter the soil. Just saying, bury all dead animals when and if an eruption happens.

[4] Okay, knowing all that. What are the odds? There are twenty of these big volcanoes globally and over 1,500 smaller ones. For the sake of simplicity, one of the twenty could, by itself, double all human annual emissions of CO_2, or even a heck of a lot more and do it in one weeklong event. As far as when, eventually. It's anybody's guess. Geologists can only apply the math they know. Which is to say, on average 50-70 volcanoes erupt every year, tiny eruptions. Some erupt every 10,000 years. Some, every 100,000 years, and some erupt every year. The point is they do erupt, and nobody really knows when. A few weeks of warning is all we'll get.

Here's my take and for what it's worth. Historically, eruptions and volcanic activity was just another bump in the road found to work against human self-domestication. That was because the number of healthy, mature forests and their unimpeded CO_2 sinks kept their emissions levels in check. So, when they occasionally get out of hand humans could still recover. As proved after Krakatoa's 536 CE and 1863 CE eruptions in comparison. Sure, many died from both, but humans as a species were able to recover and we even live without noticing smaller volcanic occurrences. However, with nowhere for volcanic CO_2 to go today, smaller eruptions or normal volcanic activity are accelerating the runaway greenhouse gas effect past the point of any possibility of human recovery after a large climate influencing eruption. Which is proven in ppm levels by the significant amount of volcanic activity that ended 130 years ago; after having occurred for most of a century. Sequestration smoothed that century out to only be a few large bumps in the self-domestication road, usually by ending short term droughts quickly.

Today, the concern specifically is volcanic activity is both expected and unpredictable. Expected, as it's part of Earth's natural order of operations. Unpredictable, because nobody knows when or where the next eruption will occur or how big it'll be. Eruptions will happen, and that certainly concerns me in the context of global warming's ppm levels of today.

Nonetheless, I'm unfortunately not done with volcanic news, not yet. I can put all this into our backyard, and into the USA. Sorry, the topic isn't the United States 1980 Mount Saint Helen's eruption. That one is a tiny VEI 4 to 5 in comparison. So, a pimple popping in comparison and only rated a VEI 5 because of the local deaths and property damage not because of its size or intensity.

MOUNT TAMBORA, INDONESIA 1815 CE

The largest and deadliest eruption in recent history, Mount Tambora, Indonesia. Another VEI 7 like 536 CE's but with less atmospheric consequences possibly due to its internal makeup. It was also possibly worsened by a significant eruption in 1814 of Mount Mayon in the Philippines. **First**, 250,000+ known deaths from the eruption alone. Which increases it's VEI rating, a lot. **Second**, what happened next took even more life.

But note, this happened before we really choked off Earth's sequestration.

The effects of this eruption helped name the event, it is known as "The Year Without a Summer." This event came before the CMS 1850ish datum. It resulted in a climate anomaly. It at first created a global volcanic winter like 536 CE but it thankfully didn't last as long. The event's emissions included acid droplets that turned the sky red. The red colored atmospheric effects lasted over ten years in the northern hemisphere and much longer along the equator.

A full year after the eruption, at the Church Family of Shakers near New Lebanon, New York, Nicholas Bennet wrote the following in May 1816, "all was froze, and the hills were barren like winter." It was snowing, as a blizzard, June 7th, and 8th in the same area. Another account from a Massachusetts historian summed up the disaster like this: "Severe frosts occurred every month; June 7th and 8th, snow fell, and it was so cold that crops were cut down, even freezing the roots. In the early autumn when corn was in the milk it was so thoroughly frozen that it never ripened and was scarcely worth harvesting. Breadstuffs were scarce and prices high and the poorer class of people were often in straits for want of food." Across the globe accounts of the disaster were recorded in much of the same way, cold, dark red sky's, and misery for the lack of food. And then the flood followed by drought.

Millions died globally from starvation and then the drought that followed; Thomas Jefferson, yeah, the founding father of the United States, Thomas Jefferson in his eighties, went into debt because of two years of crop failures to the cold. Then again to drought. It was even worse in the South Pacific and Asia, with famine lasting much longer because of the geographical result. Yes, eruptions happen all the time. It's not if it will happen, it is when it happens.

When I state that the quicker CMS is put to work for all of us, I know that I've tried to convey it with the proof required for your consideration. If you don't believe it, I can only present more facts while my hope springs eternally.

So, know this fact well; the aftermath of volcanic activity provides additional proof of sequestration science's precedence and makes CMS's findings undeniable. It all adds up.

SCOTTY, I NEED WARP DRIVE NOW!

Star Trek's Scotty character is portrayed as an engineer, so he likes math as much as I do. And so, my favorite character. They tell me not everybody likes math, which rubs against my math orientated world, but seriously, unto the majority their own. To all math haters, okay, I get it, and I do sympathize. Math is more of a profession than art so it can really suck the life out of anyone, even me. For the lack of math in this book, you can thank my wife Toni, Theo, Claire, Dave, and supplementally, Gene Roddenberry. But thanks really go to Mr. Roddenberry for creating *Star Trek* as art and not as a math textbook. An art I hopefully learned something from- don't use math in explanation, whenever possible.

Mr. Roddenberry never let Scotty explain any of the math fictionally used in *Star Trek*. Mr. Roddenberry knew nobody really wanted to hear or see it, plus it is fictional. Instead, Mr. Roddenberry's scripts filled math junkies like me with the futures depiction in poetic art. Okay, I am a geek of sorts. Anyway, all he had to do to shorten the math was to have Captain Kirk exclaiming independence from linear equations. Kirk would exclaim regularly, "Scotty, I need warp drive, now! Or energize! And my favorite catch phrase, "beam me up Scotty" to ignite the imagination of fans. Star Trek is older than I am, so I think the writing style worked. And all the calculations to repair the damaged warp core, or to transport the heroes onto a ship moving at warp were done, just in time, whoosh! Today those statements have become iconic in many realms, not just mine. That said, when I watched the latest Star Trek's, I began wondering why they needed their ship "Enterprise" if they had teleporting capabilities from one part of the galaxy to another? Seems a bit redundant, doesn't it? Just as redundant as me writing this book with a bunch of chemistry, math, and biology that nobody is going to enjoy and really doesn't want to see. Yeah, that's a good point, thanks Mr. Roddenberry.

THIS PART IS FOR THE TECHNICAL PERSON WHO DOESN'T MIND DETAILS.

A *skip* to number 15 can save time. But you'll miss out on additional information!

[1] A little element and molecule review with a plus.
   A. CO_2 atomic mass units = 44 amu. Which is heavy in comparison to other atmospheric molecules and why it acts like a blanket that covers Earth. It sinks. That is how plants gain exposure to it. because it doesn't necessarily float away like a helium filled balloon. Even though CO_2 has some buoyancy in our nitrogen, 14 amu and oxygen, 16 amu dominated atmosphere it still sinks. Which gives it the opportunity to displace lighter elements and molecules. It is not a good thing if CO_2 saturation occurs because of other atmospheric components displacement. Just something I'm thinking about and another global warming *accumulating decline*.

[2] 3.67 to 1 is the CO_2 amu to carbon 12 amu ratio found in sequestration's conversion of CO_2 into carbon 12 contained in the mass of a tree. That magic happens via photosynthesis. Which also provides the oxygen animals breathe and most carbon 12 found within fossil fuels (and our atmosphere). It is key to sequestrations ability to end global warming because it has been overlooked by many as being 1 to 1. An ignored fundamental.

[3] Carbon 12 atoms found in our atmosphere are believed to have come from fossil fuel use. But consider those carbon 12 atoms may have been recycled through photosynthetic processes numerous times and more recently than proposed. Therefore, using the form carbon atoms are in to

decide atmospheric content from emissions <u>possibly</u> becomes misleading.

A. In my hypothesis. Photosynthesis can change carbon atoms into isotopes like the ^{13}C or ^{14}C isotope. Therefore, the ^{14}C isotope dating provides no exception to the possibility it was converted to or from C_{12} or ^{13}C. Saying atmospheric CO_2 levels only come from fossil fuel use could therefore be an incomplete analysis of human and natural emissions levels. IE, some parts of emissions found in atmosphere are from global forestry and thus, the constant recycling of carbon using photosynthetic production has occurred. Furthermore, oxidation in forestry is used to form CO_2 while creating CH_4 and other gases. So, isotopes are also being formed and decayed using the kinetic isotope effect constantly. Perhaps accelerated isotope decay or production happens more often during Earth's carbon cycles than currently understood. That cycle seems obtusely intensified by today's forests' releasing far more gases than can be sequestered while regenerating forestry under a carbon hump. So, I ponder the accuracy of emissions levels solely based on atmospheric content and plan to study it further. However, none of that drastically affects CMS's current sequestration results, this is merely a question I wish to resolve and remove it from my thinking!

[4] Nonnative tree species can create major disruptions in mitigation timelines as native species out compete nonnatives or vice versa. Centuries of maturity can be lost but healthy nonnatives should not be removed without using *natural attrition harvesting* criteria. The nonnative tree's *sequestration value* is more important today to humans than being a native species is. Hey government, leave the bigger tree alone when thinning stands. Its sequestering more.

[5] *Natural attrition harvesting* is a best practice for increasing forest maturity jointly with increasing biomass production. The natural attrition process allows trees to succumb naturally while in competition for resources needed to grow. Thinning the dead and dying from forest stands improves the remaining trees sequestrations rates which also improves growth and health. The end effect promotes more natural attrition to occur while maturity thrives. The natural attrition forestry cycle is self-perpetuating without human involvement. Which results in hundreds of times more biomass efficacy by reproducing biomass faster than current immature tree management does. But it does require patience and better forest monitoring.

A. For example, a Christmas tree farm replants trees with adequate space for limbs to grow. Should that tree farm be abandoned, those trees mature. Those limbs grow because they are not overcrowded with other trees, like todays commercial replanting's. That makes a higher sequestering forest because of all those limbs making a healthier tree. All the while, the regeneration of the land around those healthy trees produces many replacement trees that thinning the dead and dying only encourages the remaining trees to become healthier faster. In addition, because the mature trees pressure the replacement trees and vice versa, biomass and sequestration production actually increase throughout the stand. Which is why the dead and dying should only be removed in rotations dependent on region and species.

B. Finally, I'm aware of very conservative landowners, loggers, and production mills that judge trees on scale value and growth rates. They'd actually prefer to leave a tree to grow then harvest it if its producing biomass. These small businesses live off the dead and dying trees and use thinning to enhance their properties remaining tree's growth rates. To these stewards of the forests, I say add maturity into your already pretty stellar planning and you'll win the trifecta. Let those big ones grow, like you do. They are worth more than any lumber value and soon they will write you a paycheck. Sequestration value will pay.

[6] Worth repeating again, a 30-year-old marketable tree absorbs around 163 Lbs. of CO_2 in 1 year, but at age 72, most species of tree used in commercial forestry can sequester 1,100 Lbs. of CO_2. Region and species are important, but all regions and all species show this potential. Some need more time than others. Location is also important. A tree in good soil outperforms one that isn't. A tree in the middle of a field with soil nutrients and ample water will have bows of sequestering limbs its entire height. One crowded by other trees or in bad soil won't and doesn't grow as well or live as long.

[7] It's thought that well over three trillion trees are on Earth. Which is a good thing, sort of. Globally, trees grow on around 9.88-10.3 billion acres globally and those lands are considered forested. But we also know they're not entirely forested nor mature forestry. Not anymore, but they used to be. What is forested or roughly 85-97% of trees (on 10.3 billion acres), are regularly logged if not protected and under 45 years old. Globally, trees could average around 20 years old because much of the global forest is in some regenerating phase and at the same time.

[8] Global forest surveys, or national ones, do not accurately document forest maturity. Instead, data displayed is to please demographics whose economic and environmental beliefs might care or not. All forestry report authors lack sequestration knowledge and that results in a lack of specific sequestration science data. Less of course CMS's recent and proven take on global warming.

[9] The *carbon hump* in forestry is a real thing because replanted or natural regeneration of trees can't absorb the CO_2 released by the harvest site waste and products produced from the harvest for 15-30 years.
   A. 60-70% of a harvested tree is waste biproduct when making typical wood products. Remember ½ of the waste's mass is carbon that turns into CO_2. If a tree weighed 10,000 Lbs. 5,000 Lbs. is carbon which turns into 18,350 Lbs. of CO_2. Biproducts are typically burned on the harvest site, left to rot, or burned to produce energy for wood product production. Therefore, replanted clear-cut forests emit massive amounts of CO_2 (carbon dioxide), CH_4 (methane), and H_2S (hydrogen sulfide) for 10-30 years. In addition, up to 80% of products made from the wood release 100% of their CO_2 over 8 to 15 years. The other 20% around 50 years. All this makes most wood products incredibly high emission sources. They are not "green" products unless they've incorporated maturity and natural attrition harvesting.
   B. Paper products are the highest emission sources in the wood industry. But I can't argue with the fact that their raw material is 100% renewable. However, that fact is easily confused with sustainable, its not. The CO_2 difference in a paper drink straw or a plastic one? There is no difference, we are *emissions dependent;* both release the same amount of CO_2 to make. One industry only lies better than the other to garner the consumers money.
   C. Not having net neutral CO_2 emissions and understanding wood and paper products impact on sequestration is by definition, understanding global warming's cause. Thus, all that green forestry stuff being spread around is propaganda. All to create misconceptions from half-truths. The truer half, trees are *binary restricted*

resources and contain *sequestration value* and that must always be accounted for, or else.

[10] **Earth is also stuck in a *carbon hump*.** Until CMS fixes sequestration, Earth is producing more emissions than it can sequester. That is a global *carbon hump* that inhibits all existing sequestration. Which is opposite to what should and can be occurring and what historically has occurred as the norm. The times this situation repeated in Earth's history brought about at least four of the seven known extinction level events. Today's global warming problem is made different only because humans can choose to adopt CMS and change how we take care of forestry resources. That avoids extinction and should be given priority, or else.

[11] **The fifth part** is proven by CO_2 ppm levels increasing since they were recorded in ice core observation and empirically from 1958 to current. Ice core records provide the observation to date all the way back to before modern humans.

   A. In addition, the blanket CO_2 applies to Earth every night does not require much of the Earth's atmospheric volume to be effective. Currently, at 0.044% of the atmosphere, it is almost 2 blankets and too hot for our comfort. Unfortunately, the number of blankets is increasing far too fast with CH_4 (methane) adding 48.75 ppm in CO_2 ppm equivalence.

   B. CO, carbon monoxide, or vehicle exhaust is less than 1 ppm. Which says something that I'm not going to say about electric cars. I'll will say they do make big cities microclimates much cleaner and people healthier.

[12] There is no longer such a thing as climate change denying humans. CMS proofs and the fifth parts relevance end all that. Unless they are brain dead zombies. So, welcome to the war on global warming, previous climate deniers.

[13] Since humans began harvesting trees we have not stopped. The forest lands on Earth have been reduced by 30% and the number of trees by 46% (due to land uses). The majority of the two declines happened within the last 250 years. Neither decline is slowing their destructive pace. This requires CMS intervention, or else.

[14] Replanted trees and commercial tree farms are expanding to meet demand. That is both good and bad. More faster growing trees are good but more immature trees that are not allowed to mature are not. The expanding of replanted lands increase *carbon humps*, they are emission sources. They add to *accumulating declines* and *constrained deforestation*. Within this realm, palm oil plantations are a very serious problem and becoming a globally shared environmental nightmare. They're replacing forest land.

[15] An interesting note: old-growth or mature tree mass can typically average 50,000 to 100,000 Lbs. Some very common tree species can regularly be many times that mass range. Yeah, many times the mass mentioned. I've personally visited several Douglas Firs estimated at over 500,000 Lbs. each! Now get this, that's their dry weights. Add in the water they hold and they're over 1,000,000 Lbs. even in today's drought. Now add this in, they are only around 140 years old! The BLM conserved them as seed trees in the 1990's when that area was harvested again. When they first logged into that area in 1960's, some 90 years ago, they were the runts of the stand and were left standing. Well, they've grown and expanded since. The BLM unknowingly did us a favor by conserving seed trees and still does.

[16] Can mature and old growth trees be added to the endangered species list? They are endangered and rare globally. Why not? They should be listed until sequestration returns a balance with emissions, to include residence levels. After that, atmospheric balance will still

need maintenance but that is not as difficult as regaining maturity is. So yes, Old-growth is an endangered species. Even though the species is plentiful their *sequestration value* as a *binary restricted resource* is not.

[17] Sequoia National Park in California has what is believed to be the largest single-stem tree on earth (by volume). A Sequoia, named General Sherman, at a whopping 4.2 million pounds, and towering at 272 feet high; it will be taller after you read this. If you've never seen it in person, I highly recommend adding it to the punch list. While you're there looking around think about this: that entire mountain range used to have General Sherman-like trees everywhere, but not anymore.

   A. Here's our crime. Before 1850, there were millions of Douglas Firs, Redwoods, Sequoias, and my favorite Sugar pines, and other tree species on the United States west coast that were taller than the General Sherman Sequioa. Yea, that should get your attention. Maybe not as massive, but much taller for sure. In example, Coastal Douglas Fir can become 330 feet, 100 meters, tall and still not be old growth. Douglas Fir can live for more than a thousand years just like the other species can. https://en.wikipedia.org/wiki/Douglas_fir

[18] Every year a tree is allowed to mature; it takes in more CO_2 from the atmosphere. It does so in an accelerated manner by applying a 3-15% per year growth rate. Exciting, right? The end of global warming even? Yes, and don't believe hype otherwise.

   A. **A fact on tree age**. If you search for data on tree growth, be careful. Many foresters say tree growth slows down after so many years. A statistical lie that is easily repeated and often cited. That statement is without thought, foresight, and highly misleading because it is half-truth. True, the growth rate percentage does slow as trees age. But it never stops growing until that tree perishes, its mass is larger and larger. So of course, the percentage decreases as the mass increases-duh. In example, I'll take the 3% growth of a 20,000 lb. slightly mature tree's mass that is 600 Lbs. over 15% growth of a 1000 lb. immature mass that is 150 Lbs. any day of the week. That is if I want a bigger tree growing more biomass or sequestration value.

B. Carbon sequestered is just under one half the dry growth obtained. Growth is proportional to *sequestration value* increasing and that can only occur with maturity. So, sequestration increases even move with maturity regardless of the smaller percents of growth that are growing a much larger mass.

C. As such, the 3.67 to 1 carbon to CO_2 ratio applied looks like this. **First**, the slightly mature tree's 300 Lbs. of carbon is 1,100 Lbs. of CO_2 absorbed! **Second**, the immature tree's 75 Lbs. of carbon is 275 Lbs. of CO_2 absorbed for the year. **Finally**, maturity wins again! Now for fun apply a 0.06% growth to the 4.2-million-pound General Sherman Sequoia. Yeah, no math is needed to make this point. However, 4,624.2 Lbs. CO_2 annually is at a minimum and probably very close to the actual. One old growth tree sequestering over 2 metric tonnes of CO_2 annually!

D. In addition, forget percentage growth rates and use logic. Even a tiny growth ring around a large diameter is much more biomass growth than a huge growth ring around a small diameter. Yeah, tiny growth rings tell us the tree is struggling, but it is still growing. Nutrients, water, species, and even soil depth can all be fixed.

E. The Titan, who is the General Sherman Sequoia, is off the sequestration scale in a good way. When you're standing by it you can feel and smell the CO_2 being converted to oxygen. It is a CO_2 scrubbing factory all by itself; the air it produces is super crisp and smells wonderful. Go see it and it's almost identical neighbors doing their important job!

[19] Fact, all tree growth rings become larger and larger every year because they must accommodate the previous year's growth ring. Trees never skip a year to grow. That's true whether you're a 500,000 Lbs. Douglas Fir, a sequoia named General Sherman, or a Maple sapling planted in your yard on Arbor Day.

[20] I'll never say "statistics" are infallible. Nefarious actors regularly manipulate statistical outcomes to generate the answer they wish to sell. I suppose they do so to promote their agenda and not our shared agendas. But as I point out, it could also be because they aren't aware of CMS or can't adjust the data received from others. To broadly answer all that, I state statistical math is infallible; if you include and show all the variables, sources, and conflicts of interest. Those trees younger than 5 years old and bare ground not appearing on a chart, trees growing less as they age, and trees can't end global warming. All misleading half-truths or just another way to openly lie and conceal facts. OR maybe there are just mistakes we've all made from our complacency to current emissions and climate change perception. Maybe, they just lack sequestration knowledge and its credibility. You decide.

[21] The Earth naturally emits 400-750 gigatonnes of CO_2 yearly, depending on the report you read. 300gt is the lowest number I've found, and I have difficulty buying only 300gt annually. Admittingly, I genuinely don't know, nobody really does. It's mostly unmeasurable; but it is estimable and is therefore a fact. Why? Even though we lack the precision to form an exact volume we are right in the assessment that natural emissions do exist, and they annually produce very high volumes of CO_2 emissions that are very likely to exceed 400 gigatonnes.

[22] Humans globally emit 35-45 gigatonnes of CO_2 yearly. Again, it depends on the report you read. And there are hundreds of these reports available. CMS applies 35 gigatonnes in all models and estimates, which I think is a solid estimate based on the law of averages and the rule of large numbers. The **rule of large numbers** I mentioned goes like this; the more educated guesses you make the better the statistical average and its accuracy becomes. So, if you educationally guess enough eventually, you'll discover the correct number in the mean value of those guesses. It does require very large numbers of samples to work.

[23] *Constrained deforestation* is widespread and affects 85-95% of global forests. It's closer to 95% than 85% and that is not a guess using the rule of large numbers, it's a fact. The 85% is the bottom of the CMS scale. As in, it is the statistically appraised and correct lower limit. Again, CMS believes the percentage to be closer to 95%. However, although the data available supports 95% CMS has yet to verify that data's integrity. However, both percents make this study's findings just as significant in outcome. Either percentage used in context of CMS's logic tells us that climate change is far worse than we thought. It also makes it impossible to argue against CMS's outcomes. What I mean is proving the impeded sequestration problem at 85% or below is made much easier to prove at 95% or above. In other words, CMS is undeniable because it can be proven, reproduced, and is useful regardless of forestry percentiles being applied. The result is verifiable at 2% or 99%. Undeniable, yes.

[24] *Unconstrained deforestation* is getting worse over time because of *demand driven forestry* practices accelerating global warming. The *accumulating decline* of global warming can and does end forests entirely. Add that the forest industry is harvesting smaller and smaller trees each

year and you begin to understand *tree degradation* occurs over time. Which is another *accumulating decline* that leads to *unconstrained deforestation*.

[25] CMS proved that global forestry fast-cycle CO_2 sequestration is not running out: it's already way out, gone, diminished, absent, lacking, and by millions of percent. <u>MILLIONS OF PERCENT</u>. Out of the entirety of all global forest, only a small percentage of all trees are mature or considered old-growth trees. The math says those trees oversee a large percentage of the CO_2 currently sequestered annually from Earth's atmosphere. Possibly 60% or more. Which is in no way shape or form providing anywhere near enough forestry sequestration. Their also damaged and being damaged by excessive CO_2 fertilization. They are lacking in numbers but are imparting their own demonstration to exactly why CO_2-driven global warming is ramping up while remaining forestry cannot do its part, lack of maturity.

[26] It is more natural for Earth to have significantly more CO_2 sequestration ability than CO_2 emissions. Plant life evolved that way to ensure their food source, CO_2, was obtainable even in low CO_2 ppm environments. Today, we have the opposite. Excessive CO_2 fertilization takes advantage of the plants ability and negatively affects plant growth by accelerating it into unsustainable proportions. This is one of the many signs of excessive CO_2 fertilization. It's empirically measurable, bad, and another *accumulating decline* from the global warming trophy humans have earned.

[27] It's estimated the ocean absorbs around 25% of Earth's human emissions only, 8.75gt. Unfortunately, that's still too much and results in acidifying Earth's oceans. Our Oceans are hitting a CO_2 self-imposed limit and can go no more without adding billions of ocean creatures that use CO_2 in calcium carbonate to make their exoskeleton

(seashells and crustations!). [ii] That is to avoid acidification because excessive CO_2 is changing the oceans chemistry while global warming increases its temperature. Acidifying is another *accumulating decline*.

[28] Less acreage is needed with more mature trees to end global warming. Which means, we have trillions of immature trees that natural attrition and *accumulating decline* kill off long before they make it over the *carbon hump*. Their death adds emissions into the *carbon hump*. All because we don't have mature trees, taking up more space to create healthy forests with the sequestration rates needed. Yikes again, I know. This also tells us that three trillion trees out there growing are actually around 700 billion trees that "might" just barely make it past a thirty-year *carbon hump* cycle before harvesting! Not near enough to bet on without CMS. It's also why replanting now can't help for half a century before scratching the surface of global warming. We don't have that kind of time. But maturity can help now if we mature a part of those 30 billion trees emerging from the carbon hump every year. That also decreases the carbon hump Earth applies towards mature and old-growth trees everywhere which provides sequestration's recovery a compounding effect.

[29] Harvesting over 15 billion trees a year and forest attrition makes those 700 billion trees over 30 years and past the hump all fall into a harvest rotation around 45-60 years of age. CMS predicted average tree age with the UN's contribution to Our World in Data and Nature Magazine articles. Except, CMS believes the tree age at harvest is up to 26 years less than the proof created with those referenceable but not precise data sources.

CRASH! IT'S NOTHING, BUT I CAN FIX IT

Climate change is a global problem. It's everyone's problem. Whether you're a newborn or not much longer for this world, it's your problem. It's not what humans want to deal with, it's what humans must deal with. And most importantly, it's a geo-engineered problem that can be reversed by CMS, or else. And it can't be passed on to the next generation, not anymore.

So far, no good. The time for or else is here and now.

If you want to know what we're in for by ignoring CMS, look at our sister planet, the long dead celestial Titan that is Venus. Long ago, it did have magnetic shielding and an atmosphere that was very similar if not identical to Earth. But it didn't last long enough to generate life as we know it. But do you know how Venus became Venus? A runaway greenhouse gas effect, just like the one Earth has today.

Volcanoes grew and spewed CO_2 that created an atmosphere so heavy it eventually crushed Venus's outer crust into a planet sized shell while it converted it's water into CO_2, acids, and eventually rock. Just like earth's volcanoes will increase as atmospheric CO_2 weight crushes our surface. Venus doesn't have continental plate drift like Earth, not anymore. Plate drift helps form a planetary sized magnetic shield that protects planets like Earth from solar radiation. Without it, solar radiation is microwaving Venus postmortem. And that heat can't leave because the CO_2 blanket Venus is wrapped in. Venus's atmosphere is almost entirely CO_2. Her average temperature is 860° F, 460°C. All thanks to CO_2. The same CO_2 we are losing Earth too.

Unfortunately, Earth's CO_2 ppm is increasing just like Venus's is assumed to have, but Earth's demise is happening in significantly less time as sequestration is still being hampered in an ever-expanding size. Fact, CO_2 sequestration is the only reason Earth did not turn into Venus millions of years ago. CO_2 breathing life evolved into plants and stopped Earth's CO_2 runaway. Again, today it's not emissions or volcanoes solely pushing Earth's ppm levels up like Venus, it is the lack and decline of sequestration creating a result that is identical to Earth's still born twin Venus.

Without plants, Venus's CO_2 levels grew into making Venus the uninhabitable nightmare it is today. The CO_2 that ended all potential for animal life on Venus is the same CO_2 we must contend with here on Earth. Yes, it took millions of years for CO_2 to transform Venus into the oven that has gone wrong nightmare it is today, but Venus didn't have humans accelerating the CO_2 problem exponentially. Unfortunately, we've created CO_2 sequestration problems that combine with our own emissions and volcano's. Yeah, we're way better at ruining a planet because we do it a lot faster.

Human activities are more efficient at screwing up a planet than Venus's planetary evolution ever was. We are changing Earth faster with forestry depredation and emissions than Venus did. Remember when I said millions of times impeded? In this scenario it's millions of times faster. Venus didn't have trillions of trees in a carbon hump emitting gigatonnes of CO_2 or billions of humans and animals releasing gigatonnes of CO_2 on its surface. Those facts accelerate the speeds of demise and are made obvious after the 1950ish datum in Figure 4-5's, page 80-83, and S1's, page 229, historical CO_2 ppm levels rising sharply above norm. Yes, it is Earth's runaway CO_2 problem. Our climate changing conditions are geo-engineering Earth into Venus faster each and every day we ignore sequestration science. Now, consider we started doing that 8,000-10,000 years ago. So, we have our history working against us along with our emissions reduction complacency today.

Look, Venus couldn't stop a destiny brought about by planetary science, but we can save Earth from a the same but greatly accelerated fate. We can own our future. CMS tells us that while providing hope.

Now is the time to ask where do we need Earths' CO_2 ppm at? Because we can now make that adjustment in humanity's benefit. We are no longer the unresponsible observer in Earth's planetary evolution. Today, we have an atmospheric control room, and we're in charge.

THE NORM IS SOMEWHERE BETWEEN 260-280 CO_2 PPM IN EARTH'S ATMOSPHERE. CALL IT 280ISH.

Is the CO_2 ppm level today, okay? I say heck no. We already know global drought and weather at this level is devastating, and so I state, "we can do much better by lowering ppm." But will we? We had better think about our sequestration, all emissions sources, and pay particular attention to volcanos before answering.

I'm first going to admit, "I don't know," and then say, "further study is warranted." Only the people living 300-12,000 years ago might know. But here's what I think I know anyways.

The 260-280ish ppm level set up around 10,000-12,000 years ago. That range seems to allow a very moderate global climate. Call it the goldilocks zone. The less than 280ish CO_2 ppm level doesn't seem to support ice ages, droughts, *accumulating decline*, or rising sea levels. 260ish ppm occurred just before nomadic humans settled into farms and started city states and might be a bit on the chilly side closer to the poles. That lower ppm limit at 260 ppm may be the best ppm level or choice. That level held up for thousands of years and just before humans began setting up *constrained deforestation's* effect. I also say it might be a satisfactory level because 536 CE events like drought still occur around 276 ppm.

The upper limit at 280ish ppm level was prior to Humanities great population expansion that happened from 5,000 CE on. That same expansion that slowly spread *demand driven forestry* practices globally and has brought us to today's 427 ppm.

What I understand for certain is today's 427 ppm is too high with nothing but global droughts, rising sea levels, and weather intensifying, and heat waves as result.

Within Figure 4's, page 80, ppm train wreck we can see what might evolve as a desired ppm level proof. This time we'll demonstrate historical ppm accounting found within NOAA ice core science.

Somewhere between 1770-1800 CE CO_2 levels were already leaving 270 CO_2 ppm behind. Then the next jump to the 1850ish datum level of 285ish ppm wasn't difficult given the sequestration damage railroads, and industrialization of forestry were creating. So, I vote for 260 ppm or even lower. How low? Maybe even 230 ppm like I used in my earlier mitigation example starting on page 161. Earth was around that ppm level for hundreds of thousands of years. That is before humans crashed the party by being smart enough to use forestry resources. We know for sure 200 ppm is too low because that's photosynthesis's lower limit, it shuts down at 200 ppm.

THE CRASH IS PREDICTED AND AVOIDABLE!

Speaking of today's 427 ppm. The issue I'm relating is CO_2 ppm doubling in less than half the time. Which causes average global temperature to increase faster and faster.

Please listen to what I say next, carefully, its important. We don't need predictions to tell us Earth is quickly becoming

Venus's twin. That happening is not a forecast it is foretold. And it really is closer than previously believed possible. We can see, feel, and measure global warmings effects, they are as real as the sun setting only to rise again. With or without us.

In context of making a prediction that I hope never happens I'm skipping the math. Earth's atmosphere achieving 840 CO_2 ppm is an indicator of terrestrial sequestration entering a point of no return. That is, a return to any climate norm that supports humans. At what point in the future does Earth's rate of CO_2 accumulation spell 840 ppm?

First, let me state emphatically that 840 ppm is the gate opening, so all is not lost at 840 ppm. Unless we continue our pace at walking off the abyss's cliff by not improving sequestration.

840-940 ppm is a range in prediction which I can explain. Earth's ppm level has been increasing each year. That rise helped show CMS's datums. For example, from 1920-1964 ppm increased an average of **0.40 ppm** per year. From 1964 to 2010 the average annual increase was **1.5** ppm per year (3x more). From 2010 up to January 2024 the increase was 2.78 ppm per year (2x more). 2023 to June 2024 the increase in ppm is an, "oh my god" 7 ppm (2x more) in a single year. The runaway? Yes, those increases say CO_2 ppm is running away. If we have 35 years before we hit 840 ppm, I'll be surprised. Add in CH_4's CO_2 equivalence, and we're already at 474.75 ppm. We're well on our way to 840 ppm if we aren't implementing sequestrations maturity curbs right now. And we're not, so far. Ahh, but if we were.

Installing CMS's maturity curbs in forestry; **First**, we'll postpone the arrival of that departure gate to a point of no return. We'll add the years between 840 ppm and us. Every growth year our scaled CMS qualified forests mature our climate changing situation improves. **Second**, that prediction, to a point, is regardless of acres and mature tree numbers we start with because we protect old-growth first. Each year they grow bigger and improve their sequestration. As they do, we'll improve our chances for humanity's climate recovery while implementing 30-year-old trees by the billions. So, all is not lost, and it can be done. That is, with a great deal of global effort it can.

Now, without CMS curbs, 840 ppm is going to come with increased temperatures. Remember CMS's point on today's 2° predictions is tomorrow's 7-10° realities. That reality currently stares at all of us from the abyss and it does so with emissions sciences and renewable energy's invitation holding hands. It also says, if 940 ppm comes it is all over for eight billion humans with very few lingering long enough to regret not improving sequestration.

Look, between 840 and 940 ppm is a point of no return because its only weeks or months after that 1,000's of ppm will register from the dead and dying of everything. Our breathable air will begin to be displaced by the heavier CO_2 molecule and all life is dead or dying from the heat, excessive CO_2 fertilization, all the other *accumulating declines*, and the biggest killer drought. Photosynthesis will stop working and with its death all hope for humans vanishes. Why? Well, today's 427 ppm is roughly two degrees hotter than 1700's 270 ppm. Unfortunately, temperature is proven not directly proportional to CO_2 ppm. Which is really unfortunate this late in global warming. Remembering from earlier that it takes half the time to double the temperature as ppm increases might help understanding the ppm to global temperature relationship.

The gist is that Earth can quickly become and has been getting geoengineered into becoming an identical twin to Venus. If humans keep ignoring sequestration 840 ppm happens sooner than we currently and incorrectly believe.

Look, it won't take hundreds of years for humans to go extinct. That might only take a few years after all the *accumulating decline* finishes off global forestry and crops when photosynthesis hits its temperature and ppm limits. It all ends fairly quickly. Yeah, within a few years of 840 ppm opening the departure gate to the abyss. Remember, nausea, severe headache and then death all occur around 1,000 ppm. At 840 nobody is going to be running any marathons.

PHOTOSYNTHESIS'S DEATH BY TEMPERATURE IS ALREADY AMONG US.

This is an *accumulating decline* poised to shut us down quicker than anything else that I can think of. Photosynthesis is very temperature dependent. Too hot or too cold and it begins to perform sub-standardly then it suspends and waits for better circumstances. The upper limit starts around 93.5° F, 34.17° C. A couple degrees hotter it suspends all activity. I don't have to tell you that it is already happening these days even in the northern hemisphere's moderate climates. Along the equator and Amazon Forests more often than not. Yes, it is getting too hot for the Amazon and African forests to sequester CO_2. Surprise, those really hot 94° F, 34.44° C days don't grow much of anything. The only that does grow is the water amounts to help the plant cool with its own evapotranspiration. They do that to keep from burning up so they can live another day. If water is not available, the end effect is obvious, it burns up and dies. I watch this happen every year when the heat waves strike. Parts of the forest around me do not get much rain, it is a high desert region. So, it turns golden brown and then heat generates lightning, without rain, and then strikes some part of it into a smoke-filled memory. That is an annually reoccurring event. It didn't used to be so often 300 years ago, but drought today has given it much stronger legs everywhere, not just the high desert.

IT'S SCARY I KNOW.

Using that 840-940 ppm spread that foretells the point of no return, life becomes unfair and unlivable. It is unfortunate for me that CMS can predict that happening just over a generation or sooner (25-35 years) (a generation is 30 years). Another top ten eruption could tip that scale in a week. It's scary I know. I was terrified when I first read the results. All I could do was look at pictures of my kids and think about their lack of a future, without sequestration. But it is a fact we must face and not one we can ignore. A truth you needed to become aware of, and now you are. And even more reason to implement CMS, if we will. I propose we do and do so at all costs. Suicide is not an option.

Look, I'm trying to encourage everyone into making the difference between CMS and falling further into the current practice of accepting global warming results with complacency. We cannot survive global warmings effects now or in the future without sequestration improvements. Nobody can. It won't matter whether you're a billionaire who plans on living for 200 years in a custom bunker or a poor kid staving from global warming's drought already and wondering if they'll make it another month. They won't and neither will you. The foretold end of sequestration is the end. Period.

SURVIVING GLOBAL WARMING WITHOUT SEQUESTRATION IS A FOOL'S ERRAND

AFTER THE GLOBAL WARMING PHASE, THE "GREENHOUSE" PHASE BEGINS.

After 940 ppm. The Greenhouse phase. This greenhouse doesn't grow plants or animals. You'd think it should by name alone. However, they're all dead and dying from all the *accumulating decline* during the phase we are currently in. Phase one, "The Global Warming phase." Then we die, just before the Greenhouse phase starts. Even ocean life is dead because photosynthesis is dead. Everything is gone. The Greenhouse phase starts. It continues to accelerate temperature increases, atmospheric pressure increases as a result, and much worse begins. This phase reforms the planet surface to it's liking. No more magnetic fields, no more air, no more surface water, and no chance for anything better than a lifeless desert.

So, let's talk about surviving climate change. A conversation that many these days are wishing to be true. That to my dismay is even being promoted as possible. **First**, there is no such thing as surviving the global warming phase. It is a fool's errand to believe otherwise. **Second**, maybe we can get away with global warming for 35 more years but not half a century. At our current levels 840-940 ppm is not that far away; accordingly, we won't last for 2 generations, not in any recognizable way. And not for very long after becoming unrecognizable either. **Third**, sorry to disappoint all those building underground bunkers for the zombie apocalypse. They won't work, there is nowhere to hide from global warmings call to death here on Earth. **Third**, literally, it's a fix it or else sort of thing because we already stand well beyond the climate changing middle ground of 280 ppm. And we're running that ground out from under our feet faster than initially believed or predicted. A lot faster. **Fourth**, okay, maybe we could live in orbit. Who wants to stare at a dead planet, breathe stale air and eat recycled food for an indefinite period? And do so while knowing the space station will only extend Humanity a decade or two at best, if that. Humans don't survive long without gravity. I do know what those people in space will regret, having passed on the actual sequestration cure. And it has shown up in time. **Finally**, even if we started building a space station big enough for 10,000 people to be self-supporting it would release enough CO_2 on Earth to finish all of us off long before it was completed. Space is not an option; not without technologies we just don't have in our toolbox.

Right now, as CMS begs for government scraps and funding, I can't help wondering about all those trillions of dollars made from forestry, emissions reductions, and the billions from government funded climate programs. That stuff made some people billions for doing absolutely nothing to end global warming and worse, some of them make it way worse. So how will they and those programs be remembered, as in they don't fix anything but riches. I also wonder how'll anything we did to end climate change prior to CMS could prove useful. That's been guaranteed by those things' results and looks as if it will be part of space life's regret. Although that regret isn't long lived it will hurt while they suffer the slow death that space offers. Good luck with that.

Literally, we have the perfect home, and they got paid to burn it into ashes, that's climate science's past, present, and future outcome without sequestration science. I'll have no part of that ideocracy or group think.

Space isn't the answer. And neither is Mars. Not even for hundreds if not thousands of years can they be. I do have my fingers crossed that someone develops warp drive, artificial gravity, solar and planetary radiation shields, ways to make food and water in space, or even hibernation technology. I can dream, can't I? The odds of doing any of that let alone all of it needed before 840-940 ppm are nonexistent. They're dreams, not reality. Surviving in space for global warmings finale is impossible. We can visit space all we want but living there is not possible. If it was possible, it would be more like winning a marathon at 105 years old than being able to walk a few blocks to elementary school when the weather is good.

And good luck with the survival shelters I've been reading about. Your gold will not protect you unless you turn it into some kind of a thermobaric suit. A CO_2-saturated atmosphere can weigh ninety times what Earth's does now. That is the same pressure as being a mile or two below the ocean's surface, yeah, that kind of pressure is not bone crushing it is bone smearing. I guess you had better build it like a deep-sea submersible, which will get a little cramped, don't invite others. Since Earth's surface became uninhabitable. Let's burrow deep into the crust or live on the bottom of the ocean. First, Earth will compress in size due to that atmospheric weight. As Earth compresses, the Earthquakes, and I mean "Earth sized quakes," generated would be so violent... not survivable, not even in a Hollywood movie. If you think volcanoes are scary now, you won't want to see the new and bigger ones forming then either. But trying to live like a rodent would have nothing to really fear. That's because we became extinct long before that crash happens. We ran out of air as CO_2 displaced it and chemical reactions from all our water and volcanic activity turned the atmosphere and oceans into acids and bases. Pure death at a planetary scale.

All the talk about surviving global warming in space, creating underground or under ocean shelters, moving further north and accepting higher temperatures is like an entire species takes part in a cult planning suicide when Haileys Comet shows back-up. My fear is those misinformed cult leaders will make it impossible to fix global warming. It is a fear propagated by current events and emissions continue to influence personal beliefs. Our perception is tainted in a very bad way.

Look, there is no hiding or surviving climate change. Global warming is like a light switch, you're either on Earth or your off, as in your extinct. Our extinction unfortunately began in 1850ish. Our end could be only 2 generations from today, maybe even sooner with a right-sized volcanic activity.

Oh, you're right to argue it won't happen overnight. And yes, you're right, mostly. I say mostly because between here and where 840 ppm shows up isn't going to be pleasant. Remember the runaway CMS discovered already has a 75-year jump on CMS mitigation. And now we also understand the global warming results we've achieved have an 8,000-year jump on our new sequestration knowledge. So, I state for the record. Without CMS, another 25-35 years and we will know more of Mother Nature's wrath than we ever cared to experience. And that can happen as "too late" becomes our only reality to share as further chaos ensues. My immediate concerns, well, I've seen them start to appear in force over this last decade and they won't stop by themselves, and neither will emission reductions do anything about it either. Literally, it might be time to wake up the neighbors and pick up the pine tar torch and pitchfork. And this time don't forget the thick hemp rope and a mule.

AND WE WERE WARNED EVEN BEFORE CMS REVEALED.

Okay, I'm somewhat older. I am Generation X. Having dated myself, I worry about the kids of today, Generation Z. They are already experiencing things like global food availability, as in shortages especially in less economically proud countries. American's see it as grocery store inflation. Gen Z also thinks these unbelievable heat waves, rising sea levels, storms coming out of nowhere and then intensifying so fast you need a stopwatch is all normal. Wars are sparking over resources, just ask Ukraine about that. Crazy and bizarre propaganda and disinformation everywhere. All the while global drought intensifies. To Gen Z those are today's norm. They did not experience the world as I did in the seventies and eighties with less drought and more abundance. A real nightmare is slowly sneaking up on them and it's progressively becoming worse, but does anyone young understand we are frogs in the pot being slowly heated to a boil? They have no memory of the pond. Are any among them willing to light the pine tar torch, sharpen the pitchfork and raise their voice? Will they use their smartphone and help spread sequestration's knowledge? Here's your chance Gen Z. You won't get another.

After two years of maturetrees.org on the internet and tens of thousands in advertisement costs along with thousands of site visits every month we have little to show in donations, members, and funding. So, I'm left to wonder; will we use it? So far, that isn't happening, and that makes human brilliance dull. So, Gen Z, you need to step it up. You are undoubtedly the first ones without any future on Earth. You and the younger generation will experience global warmings true carnage if you don't.

To help you decide, know this.

According to NASA, Svante August Arrhenius in 1896 discovered climate change. It was theorized as early as 1824. **And nobody got involved then either.** In 1938 Guy Callender proved global warming's existence. **And nobody got involved then either.** Almost a century has passed since the discovery of that inconvenient **fifth part**, CO_2 driven global warming. Now we've had five decades of emission reduction failures and industrial sized fleecing of the public. And yet, we're still working with cult-like emissions perception. Half a century of not explaining global warming cause correctly, no reproducible proof, or accurate climate change modeling. We still believe the incomplete emission's only science. And worse, we believe in emission reductions as the cure. 2019 is when CMS was first written. In 2021 openly published, and I still can't open an academic door without paying cash up front; even though they approve of my work. They can't afford being involved. Climate science is bought and paid for by the industrial mindset looking for ways to justify or create profit, not a cure.

It's no wonder belief in science and global warming is at an all-time low. All of this ceremonial dancing around the bonfire while holding the magical emissions staff has cost us time we no longer have. Oh, and don't you dare question emissions science's smoke and mirrors! I've had enough. It's time for sequestrations knotted club to tap on the emissions science cult leaders. Whether they like it or not. And you are the only one who can do that. We must gather and influence by vote and voice. There is no other way.

This book is also about how to build an ogre's club. Today, just telling others about CMS in any kind of scale costs hundreds of millions of dollars. Learning that, I didn't know what to do, I became perplexed. Still, I ignored my good sense on human behavior and the control over information these days and wrote this book. Hoping it would somehow help pay the way to spread sequestrations worthiness and knowing it wouldn't. Sorry to say, I'm realistic. But I did it anyway, while hoping people might engage and talk about sequestration with others. More dream than reality? I agree. But it is the good fight. And as we are forced to jump into the abyss we'll at least know how we got there, because Humanity was too busy to listen or act.

WE THANK YOU FOR OUR DAILY GIVENS WITH A SHINY COIN

Daily living could soon become a commodity and not the gift an overpriced grocery store provides today. Just like the 536 CE volcanic eruption that hit the reset button and changed society's directions, today we are already experiencing the onset of that exact same thing, without the mini-ice age. We all suffer from global drought whether we recognize it or not.

Thinking about how we all find it easy to avoid doing the right thing because of bad emission-based perception boggles the mind. That complacency in ideas kills. Yet, it is a very human thing, something we all do. Just ask people along the equator, oh but don't wait too long, they're starving to death so they might not be able to answer much longer.

The alternate, take the leap and warn against the wrong but popular thing and you are sure to be ostracized by the groupthink surrounding all of us. We truly are a rebellion formed by complacency to an incorrect norm. That's unfortunate. But here's CMS's phycological warning to turn our rebellion into a war against our darker side.

The global warming future within the United States continues to be the middle and upper middle classes disappearing into lower income brackets. Made possible by the economic instability created by lack of supply, inflation, increasing food costs, international market resources valuing money over human life, housing shortages due to depleted forests, and rising costs of energy created and justified by emission sciences renewable half-truths. No doubt, people will find out that just like 1812's "Year without a summer," the poorer will be in "straights for food" soon enough. This time it will be permanently sinking and without the freezing. What a nightmare for Gen Z. They really don't have a future.

Look, as annual yields in global food sources continue to decrease, they'll become only regionally sourced to support the populations growing them. No amount of fertilizer or GMO crops will change that outcome. Excessive CO_2 fertilization makes that a given. The stock market people are already betting on just that. Today, investment firms and billionaires are all buying farms, ranches, and food related industries. And driving costs up. The emission-based joke will be on them.

Farming advancements like irrigation are losing their edge fast. So, no more food exports to share with the less fortunate are in the wind here and already happening elsewhere. Self-reliance is indeed our global future as the global currency becomes food calories. Even that is certainly to be a short-lived reign. The people in the future will be willing to do anything for food, just like in 536 CE. But now we add water in.

Following today's scripture drought will keep most from producing their own food. Dry land farming will become a thing of the past. To make the point, irrigation battles are being fought in the USA's courts right now. *Look-up water wars and ground water depletion with Google.* All this will continue to evolve into much larger conflicts. Globally, they already have. I don't need to mention all that is already happening within 30° latitude of the equator.

If we hit 840 ppm food will not be the given it is today. God how I pray that I'm wrong, but CMS is not wrong and the consequence from ignoring sequestration isn't rocket science. It doesn't require a crystal ball either. It only requires my grandfather's curse. The foretold environmental damage from *accumulating decline* is just too great and happening too fast. And its measurable. Without CMS, there is no coming back from that, not without some divine intervention or some alien societies advanced technology that doesn't release CO_2. We probably should not count on either and we don't have to.

Global warming has proven much worse than 536 CE's 10-20 years of droughts. Because those droughts actually ended! That is, our global drought has been going for a century in some places. It started before 1960 CE in latitudes that span parts of the U.S.A. And now it's everywhere. Today, droughts are also worse than what people experienced in 536 CE! Remember the flood comes before the drought. Just because a region was flooded for a few weeks doesn't mean more liquid sunshine is on the way. It means it's not coming. Only irrigation and the other tools mentioned have hidden drought's impact from modernized us. If we find ourselves at 840 ppm none of our water management or farming technology will matter a whole lot. We have proved we don't need a volcanic eruption to escalate into a new norm of debauchery. We already boarded that jet and took off with a flight plan that flees straight into a mountain of CO_2 accumulations.

Have a look at what's happening to countries along the equator now. Intense storms, famine, and mass migrations, not to mention wars over food resources and poverty. All because decades of drought either produce low or more often than not, zero agricultural yields to speak of. Those areas two centuries ago were vast grassland, forests, and farms full of very peaceful people. They are no longer peaceful, they are angry. They don't like the west unless we bring food. Food is and has always been the key to peace.

We've killed the forests and the animals who maintained the Titan's existence, and our just reward is choking in the sandstorms while dying of starvation or thirst. All of this has to end right now, or else. And I'm asked good questions about that or else:

[1] When I'm asked how much time we have my answer is simple, "we don't have any time to waste."
[2] When I'm asked can we fix it my answer is simple, "yes, that is easy enough with people and funding."
[3] When asked, can we fix it in time? My answer is not so simple. It is a "maybe." I cannot answer for the volcanoes or for Mother Nature's next round of *accumulating declines* to torture our attempts with.
[4] When asked what the odds are I answer, "not great but workable, they improve every year CMS maturity curbs are constructed and held in place. There is hope."

To collectively address implementing CMS, for me personally there is only one direction to travel:

I'M GOING DOWN FIGHTING OR NOT AT ALL WITH A CMS VICTORY; I CHOOSE TO FIGHT SO I WILL WIN EITHER WAY.

VIEWING FOREST NOW

CMS's insight and logic all came as if post marked by Mother Nature herself. It occurred after I compiled a crude Microsoft Excel model that worked precisely as I'd hypothesized. In truth, it took a little practice to structure the model correctly. Anyway, that model's education sparked me to drive out into the tree covered world I live among. I enjoy wild, mountainous forested places all around my home. I can literally see four completely different ecosystems from my deck. They all had and still can grow Titans. With my new sequestration perception all of that has become far from wild and scenic, instead it became a decrepit rundown factory billowing clouds of thick tar filled smoke. That tree factory had been operating for well over a century here and longer elsewhere in the world. I then wanted, no needed to clear my head.

I got myself buckled into my old truck and drove to an acquaintance's logging site for a visit. I also wanted to look at an older logging site and some replanting efforts I donated to nearby. What I really looked for on that trip was for something to tell me CMS was wrong. I wanted comfort and not my changing thoughts by having something, really anything, tell me it just wasn't so.

Having that recently completed CMS model in my head was changing me and I did not want to be changed. So, I made the drive in an attempt to make it go away. It didn't provide a reverse of that change instead it hammered my changing perception in further. My resistance was no matter, the facts stood before me. What I viewed said it all. It was the first time I admitted just how immature the entire world's forest really is. By the time I arrived at the logging site that realization in sequestration's model and logic checked the fact box darkly. That hit me so hard it produced a few rare tears to roll off my cheek. Admittingly, because I figured out global warmings cause but more so because I could now physically see it. Knowing my past, it hurt. All because it stared back like a wrongfully beaten child with a devil's pointed tail. I could see and smell the *carbon hump's* treachery and soon realized just how screwed-up we humans really are. I just then understood that global warming was nothing like what it was supposed to be. It became an environmental impact right before my eyes as I looked across a replanted stand of immature trees. An impact we have made by converting Titans into paper towels and toilet paper. Titans were cut down to support train tracks, houses, and power lines and all of us. It was now viewable, an impact that runs deep into being human and therefore complicated to reverse. And ultimately, an impact we must reverse quickly but can't. It will require decades and even centuries of time. And most importantly, the will to resurrect Titans and save ourselves.

 I could only stare in disbelief at the endless acres of young trees, log stacks, and piled slash knowing that same scene projected itself around the entire world. And then, new climate concerns began to surface in my thoughts. At least this alternate thinking stopped forming tears. But those concerns, I still suffer from today and will likely take into my grave. Yes, it's fixable, and that would be my focus to end a grown-up's salted tears and past regrets.

 That trip finished burning all the logic and proof into my head. I could now see, touch, describe, and smell climate change. I had somberly recognized mankind's existential problem by its one and only face, a face of destruction that smiled as it lied about renewability while laughing secretly as it mentioned sustainability half-truths. Climate change was no longer conceptual or a hypothesis, it was right outside my truck's windshield, and it was everywhere. It was smellable with the windows up or down. It grew older, and then stared back at me laughing while telling me I'd found it. It was like a child dropping their hands from its eyes and saying, "you found me." But it isn't the cute attribute that makes the child adorable, it revealed the stupidity we all have by the ignorance of that child believing they disappeared. How could we have been so childlike and so dumb with our forests, so greedy? It was right there the entire time, it was definable!

    I became embarrassed for only then recognizing its abysmal form at such a late vintage of life. Recognizing it only now became an awful reckoning within an aged soul. I know it's imprint will never leave; a tattoo of an immature tree should be burned onto my forehead, justly, to advertise my shame. Right before my eyes a beauty transitioned into a sickness entrained by a much worse debilitating disease. A sight that lacked all of Humanity's best efforts of being reasonable and intelligent and one that solidified our disgrace for trying to be unaccountable. All I could see everywhere I looked were many scars from logging immaturity and another realization. We are logging ourselves towards our own extinction. We are committing environmental suicide. The scars in the hillside confirm our greedy ambitious egos willingness to run unchecked and to laugh with reckless abandonment while I just then realized a much larger tragedy awaits. We are sprinting with glee while thinking we are unaccountable for the fall ahead. That disconnect, we've falsely justified, we lie to ourselves to make it all better. But now it comes with buffering that ruins the still downloading show. A disconnect that only burns my anger into a boiling disappointment of being human because of the ignorance in thinking we can avoid the repercussions with complacency and all the uninvited distractions of propaganda. Guilt is ours and ours alone and punctuated with my remorse for once owning a chain saw and fields of timber, no longer there.

    That guilt was to place the final puzzle piece and make that road trip the worst experience of my adult life. I, at that time, seemed to be the only one in the world who knew the truth as it became self-evident. Viewing all the immature trees growing everywhere was not pictured as a beautifully lush and green mountain forest of moments past. I could now see that view only as global warmings single best effort to punish all of us eternally. We live on a planet, and we apparently love it enough to treat it badly; in the hopes it would not notice our unintentional abuse.

That plan has failed globally and did so right before my eyes. It became noticed. The scarred mountains and immature trees trying to cover for it testify to our ignorant cruelty bent towards our own destruction. I could see it all, as nobody else could or should ever have too. But you must, you must!

INSIGHT AND HALF-TRUTHS EVERYWHERE.

CMS's logic has filled me with insight. Insights that were not possible without its forestry facts. The views made possible are tainted by those ugly appearing but true results. They are in fact uniquely truthful and cannot lie, the climate change half-truth is over. The climate changing condition's truth has begun.

Sequestration science didn't lie, and it can't soften the ruined view of forests either. There is no going back now that we know the truth. All unprotected and even some protected forests that come in view now come to my disappointment. They explain exactly who is to blame, all of our demand for forest products.

It was me, not entirely, but I had done my fair share to create the carnage. I helped undo it all from balance by creating these globalized, green-covered realms of well disguised ghost's. I'd planted those little trees to conceal the harvest's deception. I'd fertilized it all with an incorrect perception about renewability and helping to end climate change. Still, all my being wanted to continue ignoring reality with complacency. But now, my addiction to profit derived from wood products could now stare back while screaming how wrong I'd been. It's like a curtain on the CMS stage was lifted while falling into blazing fragments to my feet, the act completed, the play has ended. The reveal is over. All the players bow and acknowledge the audience who victimized the play. In this play, everyone acted, so stand from your comfortable seat while applauding and take your bow with me. This ending, you and I own and if we don't use CMS we assuredly will. And then, the curtains reform and it all begins again and again. We must stop the cycle of the ignorant cheering as all of us fall into the abyss.

Then and now, I wish I didn't know a thing and had zero responsibility for CMS's realizations. You see, I'm painfully aware I am obligated to tell you about the origin of global warming, but I never wanted that responsibility. And for the information I must convey I am truly sorry. I have ruined your view of the forests, just as I did for myself and everybody else.

"Am I evil, yes I am, I am Man"-Metallica

I KNOW KNOWING SUCKS, BAD.

To show the unwanted effect, my wife and I drove through Oregon's "forested" Cascade mountains in June of 2024. Days before that trip my wife finished reading a draft of this book and provided some needed feedback. Which I won't embarrassingly relate except one, no math please. Anyway, she was aware of many parts of the years of study. However, she had not learned the entirety of the forestry picture CMS's models and logic had defined along the way.

To explain, our youngest was receiving her 2nd but related master's degree and ending, for now, her college experience. A full-time job was in her future! So, we were planning to help her move, attend the graduation ceremony and celebrations, and deliver a car to her so she could get to work, and come visit of course. That was our focus, our kid and how proud we are. But that plan became complicated during the first leg of the trip, the 278-mile road to Portland Oregon.

My wife and I were in separate vehicles for the first leg to Portland. When we stopped for lunch in Oakridge, Oregon my wife came out of the kid's car we were delivering and I from the chase vehicle. I could tell by her facial expression she was very irritated about something, but she admirably carried her discontent into the A&W for an unusual to us lunch treat, a cheeseburger, and a root beer. As we waited in line, she finally began to hint at her mood and grievance's.

She said this book was occupying her thoughts because of all the forest we were driving through. And then angerly venting, "I (CMS) had destroyed what used to be pretty." The template was cast in stone for our lunch's social topic. It would be my fault, and it is.

That early June day was beautiful, so we enjoyed our lunch treats dining outside. From there, I was able to point to the surrounding mountains and answer my wife's list of CMS questions. And right there, just as I'd written and as CMS had proved, global warming solidified in her view. It could no longer hide, not from her. From our outdoor seating I could point out the clear cuts, beetle kill, fire damage, and nothing but immature trees as far as the eye could see. While I did, I could see each of her questions battle her emission-based climate beliefs, which were brought up and finished off for good. I was amazed with sequestration science's power as it dissolved her emissions-based beliefs right before my eyes. Now something you should understand about my wife, she is one of the most intelligent people I've ever met. Her coworkers and staff will tell you the same thing or much better. Just as I will relate that she is as cursed as I, maybe even a little more.

I feel really bad for what my words did to her forest view. But I will continue to force myself to accept that and continue saying it without sugar because it really does need saying and doing. I also allowed myself a quick pat on the back for a "job done well." An arising within me that helps with the forestry pain I've placed into view, tortured myself with, and have now attached to others. That job done well contentment actually surprised me at the end of our lunch Q and A. As we left the table, she sadly commented with the now soaked in remorse view, "my god, it really is immature tree's, they're everywhere." To which I responded dumbfounded, "yes, yes they are."

SOMETHING FOR EVERYONE

When the captain of the Titanic was told there could be icebergs ahead, he thought about it but still said, "Full speed ahead, I have a record to break!" Blame it on peer pressure or whatever you want. Ultimately, he decided to ram a brand-new ship into an ancient iceberg regardless of warnings. We expect some wood companies will be more than happy to repeat that ego's mistake and gamble on the or else effect not happening.

Here's something to think about. There are far fewer and smaller icebergs in the Ocean today than there were then. Yes, another proof of the fifth part and global warming's consequences. Moving on...

CMS fully expects to deal with the captains of industry who'll be screaming, "Full speed ahead, climate change doesn't exist, wood is renewable, and humans don't need to do a thing!" In which case, they'll hit icebergs made of CO_2 (dry ice is actually made from CO_2, ironic isn't it?).

CMS also expects that after they hit the CO_2 iceberg, they'll still be screaming, "Full speed ahead!" even as the world sinks into even more climate induced chaos. But it would be better to avoid climate chaos, so let's sink their corner of the market or pass their material supply to any competitors who are willing to act on the world's behalf. At least that's what a well-funded CMS plan B does. To be sure, CMS and I don't plan on suffering idiots or the stupid who refuse to learn or find it impossible to change their mind. When I say I'll go down fighting or not at all with a win, I mean it.

My pitchfork is sharpened, my pine torch is lit, and I'm looking for iron shackles and a thick rope made of hemp. All that's left is coming up with the uber fare to get to the riot with my blowhorn. Literally, if CMS and I could just find a mature tree tall enough to hang the free riders and deniers from I would volunteer to slap the mules rump! Just kidding, or am I, no I suppose I am not. Unfortunately, it really has become that serious. I recognize the consequences for doing nothing are being made worse by fighting truth with the heaps of propaganda these days. But I must refrain because of another important fact. I must be nice to stand for CMS professionally. Thats what my wife said I must do, so I listened.

BURN THEM AT THE STAKE! BUT ONLY AS A LAST RESORT.

In these last weeks of writing, I have seen one of the largest, most profitable timber baron companies advertising how green they are with their renewable wood products. Oh wait, their all on that bandwagon. What I meant to mention is how a particular state or federal forestry agency is advertising how green they are. Oh wait, that's all of them again! Let's start over while remembering an important fact. Hardly anybody knows a darn thing about sequestration's maturity impact. They haven't done the math yet. So, all those entities are not accountable for their actions until they do. They're not mind readers nor AI driven computers, their people just like you and me. So, they have skin in this game of survival, and they need to know sequestration science warnings before we, well, skin them. That is if they don't learn anything and still spout uneducated stuff or half-truths.

"Hey, look at us, we make 'renewable' wood products," or my favorite, "We plant three trees for every one we cut down," you now know better than that, the *carbon hump* makes that just another part of *constrained deforestation*. PLEASE keep in mind, they might not. Not yet. It's more than possible that they haven't even heard of CMS. But just in case you need it, here's some talking points, or yelling in unison points. Fact-driven arguments that counters those "renewable" advertisements. And yes, this is more wishful thinking, but the following facts do stand.

First, they have to plant 3-5 saplings (trees) for every tree harvested because of the normal sapling attrition rate. Natural and human-caused attrition is hard on saplings and worse on seedlings. Of the 3-5 saplings replanted, 2-4 will die during the first year and most don't make ten years. **Second**, only one out of ten seedlings planted surviving to 25 years is likely, given region and species. Now get this, yes, those advertisements are half-truths even without CMS. The replanted trees won't replace the ones cut down until the replanted ones mature. Some majority of replants die in ten years or sooner, but a tree might reseed the land before it dies. People doing the replanting are not as good as nature at finding a good place for a tree to grow. Just replanting will never replace the trees harvested without the replants reseeding as well. All of which is post the first problem, clear cutting. Which is understood as *constrained deforestation's* effect perpetuating an immature tree problem and not improving anything but profit from *demand driven forestry*. **Third**, sequestration is not as renewable as biomass, sequestration is a *binary restricted resource*, so we'd be better off with the one tree they cut down or the forest they clear-cut. And that goes to do-gooders out there who are removing nonnative trees to improve wildlife. Please stop doing that now that you know CMS. A living thirty plus year old tree is more important than the wildlife, without it, there could be no us and no wildlife. Just plant the native trees, surround those nonnative stands with them. The native species will eventually take over by natural attrition doing its job and protecting that animal species with what we've got! So please stop making decisions based on your fundraising needs or the instant gratification a squirrel or owl might provide. If we don't mature the trees we have now, there aren't going to be any fish, squirrels, owls, or humans period. The other benefits like squirrels and owls will reproduce on their own terms, naturally if tree maturity is used in forestry decisions. **Fourth**, good forestry decisions are played out over many decades' and even centuries. Tree maturity requires that kind of foresight. Decision should be uninfluenced by human's emotional want for more instant results, instant gratification is bad in forestry. Doing otherwise is an act of *constrained deforestation* so don't act on those impulses, think forest maturity first! **Fifth**, wood is renewable, but the *carbon hump* from clear-cutting ensures its use is not a net CO_2 neutral or negative even after replanting. Again, just don't cut the tree down, thin the dead and dying instead with *natural attrition harvesting*. The older the tree stands the bigger the thinned tree becomes, don't fight the natural order. Be sure that thinning increases the stand and remaining trees *sequestration value* or don't do it. Never cut the better growing tree from a stand for economic or any other reason! **Finally**, they're perpetuating climate change by impeding sequestration. Just ask them how old that one tree they cut down was and if they expect to replant or use natural regeneration. Also, ask if it was harvested by thinning or clear-cut. Then ask when they expect the *sequestration value* to return. Well, if the trees were immature, they might not have had any *sequestration value* so leaving it, so it could mature, is the only good answer possible. All other non-*natural attrition harvesting* answers are *constrained deforestation* and *demand driven forestry*.

IT ONLY TAKES ONE.

A single mature forest can provide a greater positive impact to ending global warming than the entirety of a country building new atomic reactors, solar panels, EV's, and wind generators. No kidding, it's time to stop all this emission-based physics and chemistry defying nonsense and get to work on the actual problem, with sequestration-based science.

Just remember when thinking about the above statement, I'm not stating those things are bad pursuits in human self-domestication. They are very good things and even better for creating healthy microclimates. But they don't do one darn thing to fix global warming. Granted, they don't have the emissions that burning fossil fuels have, but they all still leak CO_2. Emissions have proven to be unavoidable, for the most part. Humans are *emissions dependent*. Therefore, *sequestration dependance* cannot be ignored!

Knowing all that and seeing those advertisements about how green any company or agency says it's car, building, or product is know this. Without sequestration science involved they can't be anything like they're advertising. I hope all this provides a new perspective on what *constrained deforestation* is. I hope this also reminds the CMS initiated of how complacency affects others and how those others need to experience the better ideas with sequestration's known impacts.

I suppose it all boils into human geo-engineering needs to be brought up to date with the knowledge of the consequences from ignoring sequestration. After all, we didn't know we were geoengineering our planet until CMS proved it with forestry. So, we also know it's time to revise our engineering, or else.

AN ELEPHANT IN THE ROOM

FIRST, AN OFFRAMP BUILT FROM TIMBER.

I await a phone call from a timber management company. I've been talking with their CEO for weeks. It could be the first full-scale CMS stewardship operation. Here's an update. "They want upfront M-O-N-E-Y, darn it! I told them CMS is just starting and has a tiny budget, okay, so they either don't care about that big burning thing in the sky coming fast, or they need the money. Either way, I'll keep trying, elsewhere if need be; maybe they'll come around eventually."

My quote to that disappointment, "it appears I need to raise coins like rabbits. And okay! All of you were right! I'll write the book for the fundraising effort! And we will go with the Cap-and-Trade model from here on out.

CMS'S REALITY AND DESPAIR ARE DISTINCTLY DIFFERENT THINGS CURRENTLY OCCUPYING THE SAME SPACE. SO...

"Timber company, here's your off-ramp to more money with biomass efficiency. Some upfront sacrifice is required. You must be willing to save the world." End of CMS advertisement.

And on to the next timber company I went, which eventually will work out. This is about finding our start, and I did write this book. What happens next? Time will tell if the rabbit market is good enough for coins to appear.

And now the elephant balancing on this book with one leg while it's trunk is erect and holding CMS skyward. Yes, it is bugling loudly because it requires attention!

Fossil Fuels. CMS is not an advocate for fossil fuels. But it would be if those producers contributed to CMS offsets. That's right, I don't care where money comes from if it can be spent on sequestration science and CMS forestry mitigation. But don't think for a second my statement doesn't come without CMS strings attached, even entangled.

CMS, and I don't advocate lowering emission standards to higher CO_2 emissions; we advocate the responsible increasing of those standards to the net zero goal. Which is made easier with CMS offsets, right now. But lowering physical emissions in all ways humanly possible is also a CMS priority. For the record, CO_2 emissions are not the only emissions generated by fossil fuel use. Therefore, fossil fuel emissions are bad and did I mention bad!

Add this, I do research and development. I'm not a specialist. As in, I'm not a PhD or well-known scientist. I'm an engineer. This work is my global climate science debut. Currently, and while I write this, I'm the guy who pays PhD's to do additional CMS research and other CMS related research I have planned. There is still lots to do because CMS's concepts and methods apply to a lot more than just forestry.

As for this writing, nobody pays or has paid me anything. So, nobody sponsors CMS's statements in any way, shape, or form. They are my own and under copyrights and patents. I'm not going to be picky about who's helping financially or what's in their past. I'm going to make them money from my efforts, if I can. The world can no longer afford the luxury of its pride by turning down an actual cure just because it requires funding from very thin air or from different thinking entities. I am a capitalist and don't intend on becoming a financial martyr for CMS's implementation. Although, as of late, that could happen. The world's fate is shared and is not my monetary responsibility alone. Plus, shared economic responsibility is the only way any CMS mitigation is going to work and perpetuate into a future without me. And if we need laws like Europe has to embrace CMS then so be it. Although, I'd prefer voluntary offset markets because we don't have time to argue.

Finally, I, the author of CMS and this book are at the time of this nonfictional writing completely independent of any industries, government, non-profits, or collegiate influences. I'm 100% independent towards everything and everybody. I don't even have a political party I associate with, but I do vote steadfastly after judging a candidates platform and never for party loyalty. I'm only at the will of facts that science pointed out and I'm left to deliver those to you without embellishment or financial incentive. To my current thoughts, all are invited to take advantage of sequestration sciences knowledge, free and openingly. The more the merrier and the faster sequestration science will begin working for everyone.

NEW IDEAS FROM THE OLD DATA'S HYMN.

It's been a struggle to find and isolate sequestration-related data simply because sequestration isn't the primary pursuit of mainstream climate science, forestry, or any related field's contemporary study, emissions are. Sequestration's effect on global warming is a new idea, a scientific breakthrough-maybe, a new discovery-possibly, or however you want to say it. I think it's a real eye opener.

ON TO DATA'S CURRENT HYMN.

Data? CMS and the maturity requirement pointed out is not the focus of public forestry data, economics and pleasing the public is. The foresters are focused on how to grow trees faster so they can harvest them sooner, even younger. Forest product engineering is looking for ways to process those younger trees. They don't know *sequestration value* or *natural attrition harvesting*, not yet.

Data? Climate science is looking for ways to reduce, measure, estimate, and predict emission's impact. Generally speaking, good stuff. However, they're studying a lot of the insignificant but sometimes very useful in effect, basically studies on previous studies. They lack knowledge of sequestration and are tied to emissions.

Data? Science in general. Studies are dictated by grants and budgets written by people who are trying to do good but can't step out of line. They and results are orientated to funding sources, their backers. They lack knowledge of sequestration and are tied to emissions. They need money and sequestrations direction to do any good.

Data? Engineering in general is looking for a technology that can't exist because if it could exist it breaks all known laws of physics, to successfully mitigate global warming. But that hasn't stopped the pursuit of a golden ticket. I also support many of these efforts. They are bold and go where most fear to tread. One cannot guess what they might accomplish, even by accident. But even the folks involved in many of these valiant efforts know they will eventually fail to the energy required releasing CO_2. But failure is learning. We need to know what not to do just as much as what we should do! It's failures hindsight that made CMS possible. So, celebrate your failures and learn. Oh, or you can use CMS to model your design and save us a lot of time, please. They lack sequestration knowledge.

Data? Most historians are studying wars and political soap operas, not forestry use. They lack sequestration knowledge and could improve CMS's historic forestry data's precision.

Data? Politicians are looking for ways to live with climate change while getting insufficient or bad data that suit agendas made from if's and but's. And of course, the alternate agenda wielding lobbyists checkbook that causes global warming. They lack sequestration knowledge and need it to inform their decisions. And then they must have the conviction to act accordingly even if it's not their party's way. Yeah, not likely.

Data? All renewable and reduced-emissions energy programs have been proven to leak carbon before, during, and past their useful life spans. They lack sequestration knowledge and need to revisit their efforts after learning it. There are a lot of good things happening in alternate energy but curing global warming isn't one of them. This entire industry seriously lacks sequestration knowledge.

Data? The one shiny spot that smells fresh surrounding the now stinking cadaver of emissions-based science's climate change is Cap-and-Trade. Cap-and-Trade can be very beneficial to curing global warming with sequestration. This data set tells of the human ability to deal with it responsibly using financial incentive. It implies tree maturity efforts can be initially funded and then perpetuated generationally. But governing its ongoing experimentation needs a little work in education. We need to teach a lot of people what caused global warming, for real. And yes! CMS's mitigation plans address that because everyone needs sequestration's knowledge.

Data? The truth in the forest industry data and other data is it's all about money. Science is plagued by money manipulating results of less than honest scientists. That will never go away when it comes to forestry or science. However, it is better for everyone's pocketbook, global warming, and the environment to firmly anchor sequestration science within Cap-and-Trade models. It should be allowed to take over those models, or else.

FIGURE S1, WHERE CLIMATE CHANGE BEGAN AND WHERE WE MIGHT END!

Figure S1 is graphed from NOAA's data from ice cores. The upward trend in CO_2 ppm began around 5,000 BCE. That is roughly 7,000 ago. That correlates with increasing human populations creating the labor pool needed to globally spread *demand driven forestry* practices. By harvesting the replacement trees sooner and sooner humans eliminated mature trees and invented *constrained deforestation* with their *demand driven forestry.* Due to the earlier harvest, on top of the earlier harvest trees have become smaller and smaller. Forests' cannot recover maturity before biomass demand harvests them again. That action removes all *sequestration value* by impeding maturity.

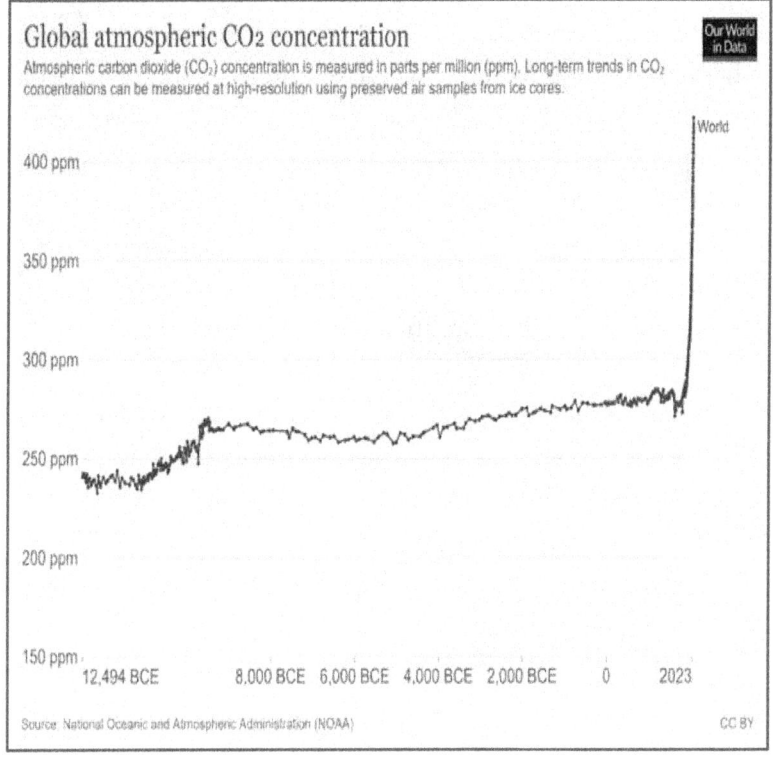

Add to that, 30% of global forest has disappeared to land use modification which resulted in a 46% decline in the number of trees. None of that happened overnight but most of it happened in the last 250 years and is well documented.

Hence, in **Figure S1** you're looking at a geo-engineered change taking place over 7-8,000 years. The impact is a result of bad forestry geo-engineering spreading immaturity and land use changes across the globe. Therefore, forestry CO_2 sequestration's global decline is both the cause and the cure. The maturity impact on forestry sequestration cannot be understated, not anymore!

Figure S1, the CO_2 ppm trends upward momentum began around 5,000 BCE (7 to 8 thousand years ago) and continues today. During this ppm rise human emissions were negligible 150-200 years ago. However, sequestration decline is highly significant and showed as such. CO_2 ppm increased as more forests were constrained by demand and their maturity was lost to harvests. Also notable, volcanic eruptions around and before the *climate changing 1850ish datum*. That datum serves as a launch point for a ppm bottle rocket like acceleration and the beginning of a sequestration-based climate collapse. This ppm trend then continues to the *1950ish datum* when the runaway CO_2 effect starts its own CO_2 fueled space rocket launch trajectory with another significant ppm acceleration.

The point of Figure S1. Thousands of years of increasing atmospheric CO_2 ppm and low human emissions until the 1800s. It all correlates historically and mathematically to human's geo-engineering our global forests. The worst of which occurred within the last 250 years at an industrialized pace and scale.

TIME TO CLOSE UP… FOR NOW

AND YES, THE BEGINNING AND END OF CLIMATE CHANGE ARE THE SAME THING, HOW COULD THEY NOT BE?

With rising CO_2 ppm over thousands of years having no correlation to emissions levels, CMS's discovery of sequestration-based correlations to ppm levels as datum points, measurable *accumulating declines*, maturity proofs, tree growth rates, CO_2 to C ratios, isotopes, all the recorded historically forestry land use modifications, millions of percent impeded CO_2 sinks, *demand driven forestry*, population expansions, the emissions tune up, forestry's very distinct correlations with ppm mitigation, CO_2 deltas proving out flow is not static, sequestration computation proving balance, all the different proofs in multiple ways, ability in forming accurate ppm predictions, volcanic demo proof, and all of CMS's logic and reproducible results. How could global warming possibly be anything other than forestry sequestration's ongoing downfall, as *constrained deforestation*?

The cure is at hand, but will we use it? I sure hope we are smart enough to fully adopt it soon enough! Global warming can only end where it all began, with us killing sequestration. That ending is with CMS's well proven intervention, by increasing forest maturity. Unfortunately, time is of the essence now that we understand global warming is a geoengineered problem and not solely an emissions reduction problem. Add that we are running out of time faster than previously believed and sequestrations involvement in the global warming battle is not just an imperative, it's required.

I've now used my grandfather's curse and warned about sequestrations demise to my own detriment as being rogue and arrogant. I feel I've lived up to my responsibility in discovery, but it isn't over yet, far from it. In addition, grandfather will be disappointed knowing I did not follow his advice and used facts to relate CMS. But I did because I had to. There was no other choice.

In closing, the only way to fix the climate mess is by addressing what we broke. We must bring back the Titans of the forest. Their sequestration is needed and a very powerful ally of the human race. Now, it is also well proven! And worthy of all of our attention, or else. We have only one job that means anything to our continued existence. Fix what we broke. Sincerely, thank you for reading this, and now the real test. Will we use it?

Subscribe at: **www.maturetrees.org**

GLOSSARY

Terms "Complete Mitigation Science" coined or expanded to describe CMS's logic.

1. **Sequestration value**

Forestry CO_2 fast cycle sinks have a *sequestration value* due to photosynthesis. This value occurs because a tree's CO_2 sequestration ability is proportional to its maturity and thus, it's mass. Thereafter, the value of sequestration becomes intrinsic to regulation of atmospheric CO_2 and as such a requirement in ongoing human self-domestication. Supplementally, in environmental history, CO_2 *sequestration value* has always been more valued than emissions in naturalized atmospheric CO_2 regulation.

Sequestration value is the second renewable resource found within trees. However, sequestration is not as renewable as a tree's more coveted and faster renewing value, biomass production. The maturity requirement of sequestration is brought about by an external *carbon hump* along with decades and even centuries required for tree growth to regenerate a useful sequestration capability. Hence, sequestration is a *binary restricted resource*.

At the tree level, increasing *sequestration value* by promoting maturity significantly increases the tree's growth potential and health. At Humanities level, *sequestration value* can generate monetization for climate regulation through cap-and-trade offsets by actualizing *sequestration dependence*. Herein, the ability to sequester CO_2 within climate regulation goals is described as a human endeavor towards the goal of further self-domestication. *Sequestration value* is currently priceless in comparison to excessive and unregulated atmospheric CO_2 levels that promise repercussions. Repercussions that do not support human self-domestication goals. Thus, making *sequestration value* a mitigating factor to global warmings origin.

2. **Binary restricted resource – an expanded definition of a renewable resource.**

Trees are considered renewable resources. Regenerated biomass from trees is only one part of the resource. Trees also have the ability to sequester CO_2 as a second renewable resource (as a terrestrial fast-cycle CO_2 sink), which is less renewable due to time required for maturity's proportional restriction with its mass that constrains photosynthetic production levels.

Trees can grow into useful biomass within two to four decades. However, to replace by growth a *sequestration valued* CO_2 sink that is significantly constrained by maturity proportionalities can require a great deal more time than its counterpart, biomass. Therefore, the sequestration resource of a tree cannot be as easily grown due to *demand driven forestry* coveting biomass more than sequestration. In contrast, a trees *sequestration value* has become more useful due to global warming impacts to further human self-domestication than its biomass.

The study's findings suggest that the CO_2 sink is more essential than the biomass among the combined resources of the tree. However, unlike its biomass counterpart, the sink resembles a nonrenewable crude oil. It can take a human lifetime for this resource within the tree to only partially be replenished by growth. Thus, though renewable, its efficacy is different from that of biomass.

The demand for biomass causes a shorter renewal duration, significantly suppressing the extended duration requirement of the CO_2 sink. Thus, sequestration is a binary resource restricted by its counterpart's popularity in commercial applications and it's slower regenerating properties. Thus, making the *binary restricted resource* sequestration a precedent for global warmings origin.

3. Unconstrained and constrained deforestation practices

Unconstrained and constrained deforestation are by CMS definition the terrestrial diseases humankind unintentionally created that allowed CO_2 and CH_4 driven climate changing conditions to form. Each result from human geo-engineering forests and relate in differing environmental impacts into climate changing conditions.

Supplementally, the scope of "deforestation" is broadened into two factors with differing roles. The assessments also oppose the conventional definition of "forest degradation," whereas forestry use is expected to result in afforestation with an expected return to a normal forestry state. And "afforestation" whereas forestry is being returned to a forestry normal.

3.1. Unconstrained deforestation

Unconstrained deforestation permanently reduces or eliminates the *binary restricted resource* of CO_2 sequestration by substituting land use by the clearing of forestry CO_2 sinks from existence. The concept behind *unconstrained deforestation* is that a forest will never return to normal forestry conditions while under human stewardship.

In contrast, "deforestation." is the removal of forests but implies afforestation as a later possibility. Thus, the problem CMS defines is related to the inherent and enduring characteristics permanently formed. In example: such as those land impacts from urbanization and a dam's reservoir construction, which will continue to exist as land use modification despite any attempts to redefine or alter their made permanent displacement of forestry.

Unconstrained deforestation is also caused by increasing climate demise in an ever-expanding spiral towards unintended land use modification brought about by *accumulating declines*. In example: The more *unconstrained deforestation* that occurs, the more climate degradation occurs, which leads to additional biological biome changes, forest fires, and adverse weather conditions, which in turn results in measurable *unconstrained deforestation* in land use and as such, ends the ability to regrow forests. In the end, all *unconstrained deforestation* factors are adjusted via climate change impacts made possible by human environmental impacts. Like deforestation, clearing forests for crop production or an unintended demise from weather, fire, climate change, biome formation, urbanization, or biological effects can be classified as *unconstrained deforestation* should that forest not keep any ability in afforestation. Thus, making *unconstrained deforestation* a precedent for global warmings origin and persistence.

The action is unconstrained in creating the destructive circumstances that perpetuate its result.

3.2. Constrained deforestation

Constrained deforestation is any practice that allows forestry to be present and growing but physically set aside to never allow or achieve a mature forestry normal. Thus, *constrained deforestation* restricts a forests ability to recover its *binary restricted resource,* sequestration. CO_2 sequestration is not currently considered in forest practices when regarding renewability. This effect is typical because of human interference with natural biology and generationally applied stewardship practices. Those practices address *demand driven forestry* first and are comprised of short-durational commercial harvest rotations that restrict tree maturity, lessen biodiversity, and increase forest impediments due to other *accumulating declines* like forest fires or drought spurred by this practice that also promotes global climate changing conditions.

This can be summarized as "the unnatural and inefficient use of forestry that impedes the capability and quantities of CO_2 sinks." *Constrained deforestation* practices also emit more CO_2 from their harvesting than the replanted or regenerative sinks can sequester and are held perpetually within a *carbon hump*. The result is owing to *demand driven forestry's* requirement for unnatural and inefficient uses. Managing forests for product demand only and not efficiently or considering both its resources is crucial to climate changing conditions and ultimately atmospheric CO_2 regulation. Thus, making *constrained deforestation* a precedent for global warmings origin and persistence.

The action is thus constrained by the destructive circumstances perpetuating their results.

4. **Climate change datums, beginning of climate-changing results.**

The datums points are CMS timeline indicators at approximately 1850ish CE and 1950ish CE. These points allow measurement starting and ending points for analysis and prediction. Their precedence is proved using the following signals, indicators, and computational analysis:

[1] Human and natural CO_2 emissions accelerating an upward trend in CO_2 atmospheric ppm levels due to a decline of available terrestrial sequestration components, IE *impeded fast cycle sinks* due to *demand driven forestry's* history of land use modification.
[2] Terrestrial fast cycle sinks abilities shrinking due to a lack of maturity that limits the volume of CO_2 that can be sequestered.
[3] Increasing atmospheric residence time of CO_2 as measured atmospherically with in and outflow deltas in their declining trend.
[4] A downward trend in the number of fast cycle CO_2 sinks available to sequester CO_2 created by *unconstrained deforestation's* finality.
[5] A correlation linked between *demand driven forestry* practices historical growth rates that increased atmospheric CO_2 ppm levels and continue to perpetuate.
[6] *Accumulating decline* factors coming into existence and then accelerating their effect on atmospheric ppm by further impeding sink quantity and quality.

The datums are made prominent within the correlation between historical forestry uses and the acceleration of atmospheric CO_2 levels recorded within ice core samples and NOAA's Manua Loa Observatory's more recent measurements from 1958-current. Thus, making *climate change datums* significant factors in defining global warmings origin and persistence.

5. Impeded fast-cycle CO2 sink or impeded sink.

Forestry or other *fast-cycle CO_2 sinks* that exist as immature sinks are a *binary restricted resource* and are impeded from reaching their required or needed potential as mature. Furthermore, impeded sinks do not hold or have only limited *sequestration value*. Typically, forestry sinks are impeded by *demand driven forestry* practices. Therefore, the impeded sinks CO_2 sequestration via photosynthesis is limited or at a nonexistent *sequestration valued* level due to the *carbon hump* and lack of maturity.

Impeded magnitudes are measured in tonnes per year of growth in CO_2 mass in the macro sense or global sense. The micro sense is measured in amu. Both are obtained in a comparison with human absence from forestry, as in the sink not being harvested previously or in the future. Impediment results of the sink should contrast with eliminating, impeding, or decreasing the volume of a CO_2 sink's micro or macro sequestering ability constrained by intervals of past, present, or projected into a future.

Impeded sequestration is also *proportional* to tree or forest's maturity and atmospheric CO_2 regulation potential. Impeded sinks are a part of *constrained deforestation* via *demand driven forestry* practices. Historically, humans impeding sinks with geo-engineering can be correlated with increased atmospheric CO_2 levels. Thus, making *impeded fast-cycle CO_2 sinks* a peer in precedent for global warmings origin and persistence.

6. Law of conservation

The *law of conservation* was used to estimate the carbon dioxide amount as being placed into an enclosed system, the same as CO_2 is present within Earth's biome.

Carbon quantities do not and cannot increase or decrease in any way within that enclosed system (Earth-bound); an increase is impossible without an external addition to the system. For instance, an asteroid hitting Earth. Therefore, Earth-bound systems producing or reducing CO_2 also conform to a law of conservation that acts as a restriction in principle. Overall, there can never be more or less elements within a closed system like Earth. Nevertheless, an element can change forms, like elemental carbon, transforming into the molecule carbon dioxide, which can then move to a different location after a chemical reaction, such as photosynthesis, converting it into biomass.

One focus of the study was the physical movement of carbon elements. The evidence suggests that the conversion of carbon through human self-domestication requires an equalizing transfer to maintain climate homeostasis, a balance acting within the law of conservation. And concludes the chemical transfer of carbon to CO_2 is inevitable and therefore unavoidable in human self-domestication efforts. As a result, conversion (emissions) and storage (sequestration) play a crucial role in achieving homeostasis within a closed biome such as Earth. The study used constraints provided by the *conservation law* as the amount of CO_2 produced by self-domestication, which is considered infinite in volume, and then placed CO_2 emissions in contrast to the deconflicting of the molecule back to carbon using forestry storage potential (as natural sequestration). The study found sequestration can quickly become finite in comparison to Emission's potential if sequestration ability is also influenced in opposition to its potential.

Thus, CO_2 accumulates in the medium between the two (emissions and sequestration). In Earth's enclosed biomes case, our atmosphere is the medium and it can become inundated or saturated when initial carbon conversion to CO_2 does not consider its conversion back into storage (via sequestration). In the case CMS's makes, should those storage abilities be hindered in any way and not exceed emissions volume the balance required by the *conservation law* (to avoid climate changing conditions) cannot be obtained. However, should reconversion storage abilities exceed emissions volumes the effect is still homeostatic because natural sequestration is historically open to applying CO_2 volumes at lower levels than its abilities and need to sequester CO_2.

Furthermore, mitigation efforts that do not involve sequestration (systems that discount the significance of storage in climate mitigation attempts or emission reduction-only attempts) are considered destined to fail at resolving climate changing conditions. They violate the law of conservation by only addressing the smaller side of the total balance needed. Therefore, emission-based attempts can never mitigate the estimated 400-750 gigatons of naturally occurring CO_2 emissions that Earth creates annually (referred to as natural emissions), Nor can those efforts address the estimated 3,600 Gigatonnes of CO_2 currently within atmospheric residence conditions that influence global climate.

And so, the *law of conservation* analysis also implies *sequestration and emissions dependency*. It is also a highly overlooked fundamental in global warming and thus the focus of sequestration computation within CMS.

7. Sequestration and emission-dependent

CO_2 emissions are defined as unavoidable in human self-domestication goals and are also the primary part of animal respiration. Therefore, humans are *dependent on emissions*. As in they must and will generate CO_2 emissions to live and improve situationally. In balance, humans are also *sequestration dependent*. As in humans must and will control sequestration of their emissions. Emissions and sequestration require balance for ongoing human self-domestication efforts in regard to our biome's requirement in continuing to support life. However, humans are more *sequestration-dependent* because of the closed system the supporting biome exists within, also defined by the *law of conservation*.

The sequestration system humans have inherited is only capable of supporting life if humans provide the stewardship sequestration requires while emissions occur as human derived and naturally occurring. Because humans have no control over naturally occurring emissions that greatly exceed human emission levels, humans become more *sequestration dependent* than *emissions dependent*.

Unbalanced interaction with CO_2 emissions allows extraneous conditions to form undesired effects on human self-domestication goals. The effects come as climate changing conditions, *associated tree and land degradation*, and the other *constrained and unconstrained deforestation* effects that generate *accumulating declines*. Humans therefore rely on emissions being balanced with sequestration. Hence, the balance is created therein. And thus, dependency is formed between the two that defines sequestration computation.

8. Application of proportionality

Evidence suggests the use of proportionality when defining CO_2 sequestration elements within the CMS studies. *Proportionality* occurs among several positions within the study and is not limited to the following examples: Note: "$\propto$," is the accepted term meaning proportional.

[1] Tree or forest maturity is $\propto$ to sequestration ability.
   A. Increasing forest maturity increases CO_2 *sequestration* ability. Those $\propto$ components are also "somewhat exponential" given enough "t" (time) to increase the forest's maturity.
[2] Maturity is then $\propto$ to sequestration ability and $\propto$ to atmospheric CO_2 ppm levels.
[3] Atmospheric CO_2 ppm residence time is $\propto$ to sequestration available in global forestry.
[4] CO_2 residence time increases $\propto$ to global sequestration abilities decreasing.
   A. The more sequestration available, the less *atmospheric* residence time.
[5] Global forestry area is also $\propto$ to maturity and $\propto$ to *atmospheric* ppm regulation.
   A. The older the forest, the less forested area *needed* to mitigate climate change when measured in atmospheric CO_2 ppm outflows created by sequestration and measured in ppm during planetary growth cycles.

9. Convenient forestry

Convenient forestry implies forestry resources that are within easy human access or control. Adjacent to human location, trail access, ample roads, railroads, low cost, lesser time required to haul resources, topographical ease to gain access, certain desired species, region, harvest method required, and the final processing location; all make some forestry locations *convenient* and others inconvenient.

In context, early nomadic people did not walk great distances or cross the horizon spanning seas just to increase their available forestry resources, like European's and Asian's did 1500 CE to current. They primarily used the *convenient forestry* and relocated when that resource or others became depleted.

Currently, all global forests have been made *convenient* in one form or another. Mostly with modernized transportation systems. The Amazon rainforest is the exception but is not immune. Encroachment into the Amazon's sequestration resource occurs daily. And as the only old-growth forests remaining at gigantic scale it is somewhat protected by zero or very limited access (that supports any kind of ground transportation). But as I mentioned that changes by the day. The Amazon forests are fortunate to be afforded both natural protections, by being inconvenient forestry, and laws like other countries national park status. Some international agreements also help protect the Amazon forests and other global forests by making them inconvenient. The notable amount of CO_2 the Amazon forests sequesters makes it Earth's last mature forest standing at a usable scale. As such, the Amazon's forests sequestration ability helps to slow CO_2 driven climate collapse with the worlds national parks and agreements really only accessorizing it's remarkable results. That is a good thing but not a wonderful thing. We need a lot more Amazon-like forests if we want to prevent and end global warming.

10. Tree and land degradation

CMS defines *tree degradation* as contributing to the forest's degradation and part of *constrained deforestation* effect on forestry biomes. *Tree degradation's* long-term effect is brought on by *demand-driven forestry* as part of *constrained forestry* practices that do not supplement soil characteristics post harvesting.

Cutting smaller trees within a reduced recovery duration diminishes tree maturity, biomass size, and the land's ability to regenerate naturally due to adverse biological effects formed from those actions. *Tree degradation* adversely affects the tree quality and, in turn, reduces the quality of wood products, precludes valuation of tree sequestration, ultimately eliminates *fast-cycle CO_2 sinks,* or furthers *impedes CO_2 fast-cycle sinks*, as it ultimately provides a biological path that pushes land and forest towards *unconstrained deforestation*'s more permanent effect.

A tree smaller at harvest is less valued. Therefore, more trees are often harvested sooner to compensate for the inefficiency in size that is lost to immaturity. In relation, often larger, and better growth performing trees are harvested during thinning to economically compensate for the smaller more predominate trees in the timber sale. Those practices are both economically, environmentally, and climate changing by promoting inefficient practices that create an unnatural occurrence biologically within tree stand soils. *Tree and land degradation* simultaneously increase with *demand-driven forestry* practices like those and perpetuate the degradation effect found with *constrained deforestation*.

For example: A 300' mature tree is harvested from virgin land / mature forest. A tree regrown in its place cannot obtain the same mass (size) as the harvested tree because nutrients from the soil are now contained within the harvested tree and have been removed from the site. Hence, the land can only support a smaller and smaller tree after each harvest thereafter. Both *tree and land degradation* occur if the soil is not renovated to sustainably support the species regrowth ability. Removing better performing trees during thinning worsens the scenario whereas the land created the better performing tree without intervention, the nutrients needed were there and now removed by harvesting the better performing tree. That action should not ever be allowed to happen.

Non renovation of soils is a widespread practice within "the natural regeneration process" so, ultimately each harvest accelerates *unconstrained deforestation* conditions to form from this *constrained forestry* result. Thus, *tree and land degradation* takes precedence in defining climate changing conditions, global warming results and its persistence, and outcome.

11. **Demand-driven forestry**

This is defined as the forestry that is managed to supply demand for forest products, disregarding the sustainability of its *binary restricted resources* or *sequestration value* as a renewable resource in general. Historically, *demand-driven forestry* generated adverse biological effects by impeding CO_2 sinks, creates *tree and land degradation,* The practice creates and perpetuates *constrained deforestation's* entirety which is this study's primary consideration in forming climate changing conditions. Therefore, such a forest that physically supplies demand only and is only managed to that extent and to no others presented is *demand driven forestry*. (7) Thus, *demand driven forestry* takes precedence in defining global warmings origin and persistence.

12. Accumulating decline

Adding to an already spiraling out of control negative result that compounds an existing problem into eventually becoming insurmountable. Also, it is commonly referred to as the snowball rolling down a hill effect.

Accumulating decline in the context of global warming are the undesired and escalating results of global warming and forestry stewardship. Undesired results brought about by human implemented climate changing conditions. Thus, *accumulating declines* takes precedence in defining global warming results, it's persistence, and outcome.

13. Natural Attrition Harvesting

Always apply maturity first to all thinning or harvesting decisions. If it does not improve maturity and forestry health don't harvest anything. Never clear-cut forest stands, no exceptions. If the tree is not dead or dying from natural attrition it should never be harvested, not until the stand can be thinned so remaining trees gain maturity, biomass efficacy, and sequestration value. Nor shall faster growth, healthier, or larger trees be selected when thinning forest stands.

Healthy nonnative trees are better than no trees due to the long-term growth cycles needed to replace their current *sequestration value*. Replant native trees only in native ways. Surround the nonnative stand with natives and allow natural attrition to remove nonnative species naturally by applying their better regional genetics. If better genes are not enough, be sure any nonnative removed sequestration value is completely replaced prior to removal. Basically, leave the nonnative alone if it's healthy.

Never ever allow demand to influence forestry thinning decisions. Forestry decisions require decades and centuries of foresight to improve maturity. Never remove a healthy tree for wildlife enhancement or any other criteria. Unless it is dead, no outer growth ring created for one growth season or less than

60% of a growth ring created the previous season, as in it is dying.

Limb production is more important to promote than height. With maturity the two become proportional because more limbs mean healthier, faster growing, and eventually taller healthier trees with higher sequestration value. This is due to limbs increasing *sequestration value* that helps the tree grow faster in good growth periods and stand strong in bad growth periods.

If the plot or stand lacks reseed potential, after it's thinned IAW *natural attrition harvesting*, immediately replant seed trees so it can evolve regeneration ability to improve longevity, maturity, and biomass production. Always use natives when replanting, whenever possible.

14. The Carbon Hump (1)

The *carbon hump* describes a required relationship formed between carbon sequestration and carbon emissions found within the *law of conservation* when applied to carbon in atmosphere and carbon holding biomes. Carbon emissions exceeding a balance with carbon sequestration creates excesses that sequestration must overcome to regain that balance. It is referred to in both the micro and macro senses.

AUTHOR BIO
Timothy C Thompson

The short version of my bio; well, that's easily summarized. Raised free range as a middle child after being born in less than wild conditions. Thanks to public school teachers, I survived the peer pressure, predators, and eventually escaped the repeat offenders in a flock of wild siblings. I survived again, this time, the politician-driven Army's attempt to find a place for me to die. But I found wisdom and education from the better Army experiences. Eventually, the young Army man shaped into loving academics and beginning a family of his own. So now as an adult, I'm a father who's all grown-up after being raised by his wife and children. Now I'm living the outdoor dream of an Oregonian. A scientist gone rogue from emissions. A man who believes he is righteous in scientific words, acts, and naïve enough to think people will actually listen. A no-nonsense nonfictional author who likes reading and working undisturbed. An ogre sized man with a matching past and one with a newly formed knotted club and shield to wield. I do need therapy because intervention has failed to remove me from an obsession with science. Hmm, maybe creating a fictional story could be a self-prescribed therapy for having to share the nonfictional sequestration news I forced upon me? No, I don't have time to externalize wishes and dreams. Dealing with global warming is far too demanding, there is so much still to do.

The long version, I'm situated within the spread of eight children in the upper middle, and the oldest of four boys. My parents are made from multiple divorces, remarriages, and continued breeding habits. Being the middle child with problematic siblings as the book ends did have advantages. I could leave for days and neither of the parental guardians nor the book ends took notice. The multiplicity of parental figures only greased the squeaky wheels and never the smoothly running hubs in the middle. I suppose they didn't have to. What is the long-term effect? I'd like to think I'm self-starting, self-sufficient, and a good parent, but I'm also biased in family-based opinions.

Raised mostly in the backcountry of Montana. At one point, I attended a school slightly larger than one room. My education, mostly, came from very diminutive resources that included hundreds of past due public and school library books with Scholastic Book fairs as an occasional carrot. I am a voracious reader and obsessive.

I was never a model student, possibly because of all the moving and instability in my youth; but I don't believe that. Anyways, ten different schools that I remember can be associated with my attendance, before three high schools in three states. All those schools devoted to my youth in some way and believe me, I wouldn't change a thing; although, I, like everyone else, can't help wondering what might have been. But also, like most, I'm over it.

From the time I was twelve I've worked, first by lying about my age and then for myself. Doing so seemed obvious since that seemed my place, a worker bee meant to serve the hive. But along the way, somehow, memorable teachers reached out and pointed me towards something else, something I would have to reconcile to become an adult, undisciplined me.

Public school teachers. Not all of them disliked my filthy, non-ironed, and ragged sighting enough to not say "Tim" you are going to do great! A few sacrificed their daily norms and influenced me positively. They showed compassion even though they were not paid to do so. A few of them wasted enough of their time with the new kid to make the positive, life-long impressions this kid needed. Those impressions accumulated.

It was somewhere in one of the many middle schools a science teacher recommended that I begin keeping a journal. It allowed me to record ideas and concepts to research later. But I think he did it so I would stop asking so many questions. Plus, it made me write. But I'll never claim superior writing ability, that I am not. However, keeping the journals did force me to become experienced in problem-solving and conducting research. Today, at my current vintage, the ideas do not come as fast but the journals I've kept still produce stuff I love to work on. I do admit most of my journal topics perish to experience or research, but not all have. That is especially true for my climate and forestry work. When your right your right. But I'm sure my Mars reentry vehicle is a failure. Although, I did get an award from NASA for a rover project. Anyway, it was somewhere in my senior year of high school that all those good teachers and their impressions coupled into a shockingly good SAT score. Yes, I studied for it. Even scraping enough money to buy a good SAT guidebook. The problem was, it was late in the teenage game, and I had no hope whatsoever to pay for college nor the grades for academic scholarship. So, I did what anybody trying to break out of being limited in a social caste, being poor that is. I took the only desperate choice available.

I took the required tests, did surprisingly well again, picked my job from the list, and volunteered in the U.S. Army. That took months for me to decide. Meanwhile, I was bagging groceries, repairing cars, tossing hay bales, and cutting grass for money. But I finally put down my baseball-equipped dreams of youth, a group of good friends, and joined. Man did I like it. Three meals a day, exercise, and very interesting people to work around. Oh sure, it had many downsides like the remarkably interesting people to work around; but my first years in the military were fantastic, but I was a naive. But all good things end as maturity continues to develop. So as many global calamities and political disappointments continued to kill the hapless for being ignorant, I decided to opt out of taking part. I accepted that naive politicians and not the Army would end my military career. So, after fulfilling my ten-year agreement, honorably discharged, and empowered to enter a new vocation I moved along, and happily left aviation behind.

My wife and kids and my first professional job all stuck well to my campaign of achieving adulthood here in Oregon. I guess it was in Central Texas and during college that all started to go well. For several years after the Army, I spent the highly inadequate GI bill, taught non-credited classes, tutored, attended classes, worked for myself. I managed a four point oh academic scholarship and then a government internship. So, after student loans and my lovely spouse helping support me and the kids I earned my little degree in engineering with honors. Which eventually moved me forward to here and now, doing what I discovered as what I'm supposed to do. Provide a better world than given, hopefully.

Thanks for reading, it really does mean a lot to just be heard and hopefully, be taken seriously.

www.maturetrees.org has more information.
Please make the difference by subscribing.

BIBLIOGRAPHY

Thanks to the published contributions below. The following are a few of the papers cited within the CMS manuscripts' and mentioned in Resurrect Titans

---

[i] Sequestration data complements of US Department of Energy, Method for calculating Carbon Sequestration by Trees in Urban and Suburban Areas, April 1998

[ii] Jiang, LQ., Kozyr, A., Relph, J.M. et al. The Ocean Carbon and Acidification Data System. Sci Data 10, 136 (2023). https://doi.org/10.1038/s41597-023-02042-0

1. **Land use strategis to mitigate climate change in carbon dense temperate forests.** Law, Beverly E at al. (2018): 3663-3668, s.l. : Proceedings of the National Academy of Sciences of the United States of America, Vols. 115,14 . doi: 10.1073/pnas. 1720064115.
2. **Our World in Data, NOAA/ESRL.** Global CO2 atmospheric concentration. *Our World In Data.* [Online] 06 12, 2022. [Cited: 06 12, 2022.] https://ourworldindata.org/explorers/climate-change.
3. **Our World Data.** Global CO2 emission from Fossil Fuels. [Online] Our World Data.org, Dec 12, 2022. https://ourworldindata.org/co2-emissions.
4. **U.S. Environmental Protection Agency.** *Greenhouse Gas Mitigation Potential in U.S. Forestry and Agriculture.* Washington : EPA, 2005.
5. **U.S. Department of Energy.** *Method for Calculating Carbon Sequestration by Trees in Urban and Surburban Settings.* Washington, DC : U.S Department of Energy, 1998.
6. **Food and Agriculture Organization of the United Nations.** Proportion of forest area within legally established protected areas, %. *2000-2020.* s.l. : United Nations Council.
7. **By Louise Walsh.** Silent Witness, Tree Rings 536 CE. *Cambrige University.* [Online] Cambrige University. [Cited: October 3, 2023.] https://www.cam.ac.uk/silentwitnesses.

8. **National Association of State Foresters.** Timber Assurance. *National Association of State Foresters.* [Online] Dec 2021, 2021. [Cited: Febuary 20, 2022.] https://www.stateforesters.org/timber-assurance/legality/forest-ownership-statistics/.

9. *U.S. Forest Resource Facts and Historical Trends FS1035.* United States Department of Agriculture, Forest Service. Washington DC : United States Department of Agriculture Forest Service, 2014. FS-1035.

10. **Hammerschlag LLC.** *Uncaptured Biogenic Emissions of BECCS Fueled by Forestry Feedstocks.* On Line : Hammerschlag, NR-040(g), 2021.

11. **Forests: Carbon sequestration, biomass energy, or both? Alice Favero, Adam Daigneault and Brent Sohngen.** 13, s.l. : Amaerican Association for the advancement of Science, Januarary 2020, Vol. 6.

12. **Andrew Moseman, Daniel Rothman.** How much carbon dioxide does the Earth naturally absorb? *MIT Climate Portal.* [Online] MIT, January 4, 2022. [Cited: August 27, 2023.] https://climate.mit.edu/ask-mit/how-much-carbon-dioxide-does-earth-naturally-absorb.

13. **Birdsey, Richard A.** *Carbon Storage and Accumulation in United States Forest Ecosystems.* Radnor : United States Department of Agriculture, Forest Service, 1992. WO-59.

14. **Brink.** How to calculate the amount of CO2 sequestered in a tree per year. *https:/www.UNM.EDU.* [Online] University of UNM, 2021. [Cited: November 15, 2021.] https://www.unm.edu/~jbrink/365/Documents/Calculating_tree_carbon.pdf.

15. **Hygromechanical Mechanisms of wood cell wall revealed by molecular modeling and mixture rule analysis.** Chi Zhang, Mingyang Chen, Sinan Keten, Benoit Coasne, Dominique Derome and Jan Carmeliet. 37, s.l. : Amaerican Association for the Advancement of Science, Sept 2021, Vol. 7.

16. **Ciais, P., C. Sabine, G. Bala, L. Bopp, V. Brovkin, J. Canadell, A. Chhabra, R. DeFries, J. Galloway, M. Heimann, C. Jones, C. Le Quéré, R.B. Myneni, S. Piao and P. Thornton ,.** *Carbon and Other Biogeochemical Cycles.* Cambridge, UK and New York : Cambridge University Press, 2013.

17. **Coder, Dr. Kim D.** *Trees Per Acre Table.* s.l. : University of Georgia, 1996.

18. **Cornwall, Warren.** The Burning Question. *Science.* Science, January 2017, Vol. 355, 6320, pp. 18-21.
19. **Crowther, T. W., Glick, H. B., Covey, K. R., Bettigole, C., Maynard, D. S., Thomas, S. M., .. & Tuanmu, M. N.** Mapping tree density at a global scale. *Our world in data,* . [Online] Nature, October 13th, 2020. [Cited: 9 12, 2023.] Linkhttps://www.nature.com/articles/nature14967. 525(7568), 201-205.
20. **Jean-Francois Bastin, Yelena Finegold, Claude Garcia, Danilo Mollicone, Marcelo Rezende, Devin Routh, Constantin M. Zohner and Thomas W. Crowther.** The Global Tree Restoration Potential. *Science.* July 2019, Vol. 365, 6448, pp. 24,76-79.
21. **Keenan, T.F, Williams, C.A.** *The Terrestrial Carbon Sink.* Berkeley, Worchester : Annual Review of Environment and Resources, 2018. 102017-030204.
22. **The Generalized Chapman-Richards Function and Applications to Tree and Stand Growth.** LIU Zhao-gang, LI Feng-ri. Harbin 150040, P.R. China : Journal of Forestry Research, 2003, pp. 19-26. 1007-662x(2003)01-0019-07.
23. **Land-use change and carbon sinks: Econometric estimation of the carbon sequestration supply function.** Lubowski, Ruben N., Plantinga, Andrew J., Stavins. [ed.] Science Direct. s.l. : Journal of Enviromental Economics and Management 51, 2006, Journal of Environmental Economics and Management 51, Vol. 51, pp. 135-152.
24. **MacCleery, Douglas W.** *American Forests A History of Resiliency and Recovery.* Durham : The Forrest History Society, 2011. ISBN 0-89030-048-8.
25. **Historical and recent changes in the Spanish forests: A socio-economic process.** María Valbuena-Carabaña a, Unai López de Heredia a, Pablo Fuentes-Utrilla a,. s.l. : Science Direct, 2010, Review of Palaeobotany and Palynology, Vol. 162, pp. 492-506.
26. **Melby, Patrick.** Insatiable Shipyards: The Impact of the Royal Navy on the World's Forests, 1200-1850. *https://www.wou.edu.* [Online] 2012. [Cited: December 3, 2021.] https://wou.edu/history/files/2015/08/Melby-Patrick.pdf. HST 499.
27. **Oosthoek, Jan K.** The Role of Wood in History. *https://www.eh-resources.org/the-role-of-wood-in-world-*

history/. [Online] ENVIRONMENTAL HISTORY RESOURCES, 1998. [Cited: Febuary 17, 2022.] https://www.eh-resources.org/the-role-of-wood-in-world-history/.

28. **Oosthoek, K.Jan.** The Role of Wood in History. *Environmental History Resources, eh-resources.org.* [Online] Environmental History Resources, Febuary 7, 2022. https://www.eh-resources.org/the-role-of-wood-in-world-history/.

29. **Petersen, Georgia.** *How Much Lumber in That Tree?* s.l. : Michigan State University Extension, 2004. E-2915.

30. **Rafferty, J. P. and Jackson, . Stephen T.** Estimated Temperture variations for northern hemisphere and central England (1000-2000 ce). *Britannica.com/science/Little-Ice-Age.* [Online] Encyclopaedia Britannica, Inc. https://www.britannica.com/science/Little-Ice-Age.

31. **Serget Paltsev, C. Adam Schlosser, Henery Chen, Xiang Gao, Angelo Gurgel, Henery Jacoby, Jennifer Morris, Ronald Prinn, Andrei Sokolov, Kenneth Strzepek.** *2021 Global Change Outlook.* Cambridge : Massachusetts Institute of Technology, MIT, 2021.

32. **Smith, James E., Heath, Linda S., Skog, Kenneth E., Birdsey, Richard A.** *Methods for Calculating Forest Ecosystems and Harvested Carbon With Standard Estimates for Forest Types of the United States.* Newtown Square, PA : USDA Forest Service, April, 2006. General Technical.

33. **Steele, Philip H.** *Factors Determining Lumber Recovery in Sawmilling.* Madison : USDA, Forest Service, Forest Products Laboratory, 1964. FPL-39.

34. **Thompson, Timothy C.** *Apparatus, System and Method for Construction of Buildings and Structures, E3Lumber. 17/145,367 application* United States, January 19, 2021. Utility Application.

35. **Vanhorn, J.** Approximate Weights of Wood in LBS.PDF. *www.oocities.org.* [Online] 2004. [Cited: November 23, 2021.]

36. **Wood, Smith and.** *History of Yard Lumber Size Standards.* Madison WI : Forest Service, United States Department of Agriculture, 1964.

37. **Y., Yan.** *Integrate Carbon Dynamic Models in Analyzing Carbon Sequestion Impact of Forest Biomass Harvest.* EPub. EPub : SCI Total Eviron, Feb 2018. 15;615:581-587.

38. **A sustainable wood biorefinery for low–carbon footprint chemicals production**. Yuhe Liao, Steven-Friso Koelewijn, Gil Van den Bossche, Joost Van Aelst, Sander Van den Bosch, Tom Renders, Kranti Navare, Thomas Nicolaï, Korneel Van Aelst, Maarten Maesen, Hironori Matsushima, Johan M. Thevelein, Karel Van Acker, Bert Lagrain, Danny Ver. 6484, s.l. : Amaerican Association for the Advancement of Science, March 2020, Vol. 367.

www.ingramcontent.com/pod-product-compliance
Lightning Source LLC
Chambersburg PA
CBHW060455030426
42337CB00015B/1598